TOWARDS FOREST SUSTAINABILITY

Edited by
David B. Lindenmayer & Jerry F. Franklin

Island Press
Washington • Covelo • London

05- 348

Library of Congress Cataloging-in-Publication Data

Lindenmayer, David.
Towards forest sustainability / David Lindenmayer and Jerry Franklin.
   p. cm.
Includes bibliographical references (p.   ).
ISBN 1-55963-381-6 (pbk. : alk. paper)
1. Sustainable forestry. I. Franklin, Jerry F. II. Title.
   SD387.S87L56 2003
   634.9'2—dc21
                    2003007824

Published exclusively in North America by Island Press
1718 Connecticut Avenue, N. W.
Suite 300
Washington, D.C. 20009
www.islandpress.org

Available worldwide (excluding North America) from:
CSIRO PUBLISHING
150 Oxford Street (PO Box 1139)
Collingwood Vic. 3066
Australia
www.publish.csiro.au

Cover design by Jo Birtchnell
Text design by James Kelly
Set in Stempel Schneidler 10/12.5
Typeset by James Kelly
Printed in Australia by Ligare
Front cover photograph courtesy CSIRO

Disclaimer
While the author, publisher and others responsible for this publication have taken all appropriate care to ensure the accuracy of its contents, no liability is accepted for any loss or damage arising from or incurred as a result of any reliance on the information provided in this publication.

Contents

short term. Nevertheless, major changes in native forest management are occurring in British Columbia, the United States, Scandinavia and Australia to better conserve key ecological values. There are important common factors in the changes that are occurring, despite the specific differences among temperate forest nations. Globalisation of the world economy has created an over-arching—and rapidly changing—context since these countries are all heavily involved in trade in wood and wood products. As well, there are many commonalties among native temperate forests in their natural history and in historical treatments of these forests.

These commonalties led us to believe that sharing ideas and experiences of transitions in native forest policy and management would be useful. Hence, a roundtable meeting on forest management and conservation was convened in August 2002 at Marysville, in Victoria. The meeting was hosted by the Centre for Resource and Environmental Studies (The Australian National University) with funding provided by The Myer Foundation and the Poola Foundation. Attendance included experts in forest economics and policy, water management, and timber production as well as ecologists and conservation professionals.

General issues associated with management and conservation of native temperate forests—and specific issues associated with the montane ash (*Eucalyptus* spp.) forests of southeastern Australia—were discussed at the Forestry Roundtable Meeting. The Victorian montane ash forests represent one of the most productive and economically valuable forest areas in Australia, but they are also forest ecosystems that have extraordinary values for conservation of biodiversity and as source of domestic water supplies; hence, substantial conflicts have arisen among forest stakeholders. Representatives from Victorian state government agencies responsible for forest conservation and water management, as well as key forest industry and conservation groups, were participants in what proved to be stimulating and productive discussions.

Some of the participants in the roundtable meeting were invited to write essays on the topic of transitions to sustainability in forest conservation and management, and these essays form the chapters of this book. Collectively, the essays span the full range of spatial scales from regional and state through national to global perspectives on ecologically sustainable forest management.

Underlying these discussions is the reality that sustainability is not a set endpoint but, rather, an overall direction in conservation and forest management that reflects an evolution in societal perspectives and scientific knowledge. We hope that this book, which incorporates the considerable experience of the contributors, will help to propel forest management on its 'sustainability pathway' into the future.

David Lindenmayer and Jerry Franklin (Editors)

Participants at the Forestry Roundtable Meeting

Professor Mark Adams	University of Melbourne
Professor Per Angelstam	Swedish University of Agricultural Sciences
Melissa Boyd	OfforSharp
Dr Mick Brown	Forestry Tasmania
Professor Fred Bunnell	University of British Columbia
Dr Rob Campbell	Department of Sustainability & Environment
Dr Judy Clark	Centre for Resource & Environmental Studies, Australian National University
Associate-Professor Stephen Dovers	Centre for Resource & Environmental Studies, Australian National University
Tim Fischer	Australian Conservation Foundation
Professor Jerry Franklin	University of Washington
Kertsen Gentle	Timber Communities Australia
Graeme Gooding	Victorian Association of Forest Industries
Lindsey Hesketh	Australian Conservation Foundation
Dr John Hickey	Forestry Tasmania
Professor Aila Keto	University of Queensland
Doctor Laurie Kremsater	University of British Columbia
Dr Charles Lane	The Myer Foundation
Frank Lawless	Melbourne Water
Associate-Professor David Lindenmayer	Centre for Resource & Environmental Studies, Australian National University
Ian Miles	Department of Sustainability & Environment
Professor Jari Niemela	University of Helsinki
Associate-Professor David Norton	University of Canterbury
Tim Offor	OfforSharp (roundtable facilitator)
Professor David Perry	University of Oregon
Professor Michael Soulé	University of California
Lady Southey AM	The Myer Foundation
Malcolm Tonkin	Hancock Victoria Plantations
Dr Sally Troy	Parks Victoria
Brian Walters SC	Flagstaff Chambers
Marcus Ward	National Forest Summit
Professor Bob Wasson	Centre for Resource & Environmental Studies, Australian National University
Mark Wootton	The Poola Foundation
Virginia Young	The Wilderness Society
Dr Lu Zhang	CSIRO

Acknowledgments

The Forestry Roundtable Meeting held in August 2002 at Marysville, Victoria, Australia, from which this book flowed, was made possible only because of the generous support of The Myer Foundation and the Poola Foundation. In particular, Charles Lane and Mark Wootton helped to keep the process on the rails. Bob Wasson, Director, Centre for Resource and Environmental Studies, provided considerable guidance in the nuances of successful roundtable discussions based on his extensive experience in the Australian water and cotton industry sectors. Ian Miles from the Victorian Department of Sustainability and Environment made a major contribution to the smooth running of the field day associated with the Roundtable Meeting.

The many participants at the Forestry Roundtable Meeting contributed extensively to very informative and useful discussions, and their ideas helped stimulate many of the authors in writing their essays to reshape the material in the chapters contained in this book.

Finally, we are most grateful to Nick Alexander from CSIRO Publishing for his assistance in making this book a reality.

1

Challenges to temperate forest stewardship—focusing on the future

Jerry F. Franklin

Too many participants in the current forest policy debates—stakeholders, media, politicians, and resource professionals—appear focused on the past rather than on the future. Major economic and social changes are dramatically altering the context within which forest policy will be created and implemented. Equally important are shifts in the types and intensities of challenges that we face in sustaining critical forest functions—such as protection of watersheds—and forest biodiversity.

In North America and many other regions of the world, stakeholders and politicians continue to fight the resource war of the 20th century—preservation versus exploitation. These battles are familiar and comfortable. But the major challenges of the 21st century are not likely to repeat those from the previous century. Few of the forest policy debates, alliances, and 'solutions' of the 20th century are likely to be relevant to these new challenges.

Society—by continuing to focus on old issues—is also failing to recognise the fundamental changes that are occurring our interrelationships with native forests, including the human stewardship necessary to assure the continued health and functioning of native forests, and their declining role as sources of wood products. In this essay I share my perspectives on some of the important circumstances and issues regarding forest resources that national and global societies must address in the 21st century. The focus is primarily on temperate regions and the developed world. Because of the variety of social and economic factors, the challenges of tropical forest policies are currently and likely to continue to be very different from those affecting temperate and boreal regions.

Globalisation of the wood products industry

Globalisation of the wood products industry is, I believe, the most significant factor influencing the developing context for forest stewardship. Technology and globalisation of the world marketplaces are creating a new model for production of the bulk fibre needed to provide for the mass markets in wood products, such as paper and common wood construction materials.

The global marketplace, with its emphasis on moving activities to areas where lowest per unit cost of production can be achieved, fits perfectly with basic corporate goals. Consolidation also appears to fit the global paradigm so we can expect that a few very large, international, publicly traded corporations will increasingly dominate global markets in wood products. The goal of these corporations is to maximise their return on investment; hence, the net present value model will continue to dictate corporate forestry practices just as it has for at least 50 years. Net present value is the discounted values of the revenues and costs from the use of forest resources over time and allows investors to compare returns from different forest resource investments as well as with returns from alternative investments. Maximising net present value is, consquently, taken as the overall goal for forest management under this model.

Technological developments in forestry—as well as the fortuitous discovery of some unusual biological potentials—fit very well with the goals of global corporate forestry. First, there has been the recognition of the extraordinary production potential of some exotic tree species—initially, some pines and later (with development of appropriate pulping processes) eucalyptus species—on temperate, subtropical, and some tropical sites, primarily in the southern hemisphere. Second, there has been the increasing potential for enhanced production using bio-engineered trees—

initially by traditional breeding methods and more recently by direct manipulation of genes.

As a result of these developments, corporate forestry is moving to an agronomic model of wood fibre production and away from traditional forestry models. In effect, the move is towards the development of fibre farms in which intensively managed 'fields' of engineered poplars, exotic pines, and eucalypts are grown on short rotations of two or three decades—perhaps eventually in only a single decade. Adoption of the agronomic model has many advantages from the corporate perspective including much shorter investment cycles, high efficiency in per unit production costs, and reduced environmental constraints, since societies often have fewer expectations of farms—including fibre farms—than they have of areas that are clearly forests.

Temperate and subtropical regions of the southern hemisphere are primary sites for the development of corporate forestry's fibre farms for many reasons.[1] Locations include Australia, New Zealand, Chile, Argentina, Uruguay, southeastern Brazil, and South Africa. Productivity of exotic species is extraordinary on many southern hemisphere sites, often at the very upper limits of known wood production. Many of the sites have been previously used for agriculture or grazing so costs of preparing sites or mitigating environmental impacts is often low. The ability to use efficient equipment for tending and harvesting the plantations substantially reduces labour costs. Furthermore, labour and other production costs, including those related to environmental concerns, are typically lower in many countries at lower and southern latitudes than in North America and Europe.

The implications of these new opportunities for corporate investment in fibre production are inevitable. Few, if any, forest sites in the northern hemisphere can compete directly with fibre farms in the southern hemisphere even in terms of biological productivity, let alone total cost of production. Only the very best of the forest sites in the northern hemisphere can approach the productivity levels of exotic plantations in the southern hemisphere. When you add in the costs of labour, taxes, and environmental mediation in North America and Europe the advantages of

1 See Marchak, 1995

investing in agronomic wood fibre production in the southern hemisphere are very large. Indeed, the only advantage of north temperate forests appears to be in their proximity to large markets; unfortunately, transportation costs for finished products are typically not a high proportion of the total cost.

Hence, following trends in global production and marketing of the last several decades we can expect that corporate wood production is going to continue to move (1) to the agronomic model and (2) primarily to southern latitudes. It is inevitable based on the imperative of maximising return on investment. Please note that this shift is not about exporting North American or world demand for wood fibre to 'third world countries' that 'do not know how to manage their forests and have no environmental laws or regulations' as suggested by some observers. Also, generally we are not talking about cutting down native forests in order to provide space for exotic plantations, although this has occurred in the past. The majority of thse plantations are on abandoned agricultural or grazing lands. Furthermore, these developments are not likely to significantly affect the fate of tropical forests. Over 85 per cent of the tropical tree harvest used for wood products is used in-country and does not enter world trade; the only significant global market for such wood is in environmental rogue states, such as Japan, China, Taiwan, and South Korea. These shifts in the locale and techniques for the wood products industry *are* the consequence of corporate decision-making based on standard business models and practices. These current developments have made a major contribution to the current glut of wood fibre in the global marketplace. They are likely to continue to provide excesssive amouts of wood, at least until the consequences of the shift to fibre farming are fully recognised and appropriate adjustments are made. We can expect that there will be a significant competitive 'sorting out' among countries and regions that are heavily invested in fibre farming; factors affecting the outcome will include relative productive capabilities and total costs of production. Substantial improvements in productive capacity and reductions in costs of production can be expected in subsequent generations of fibre farming as a result of genetic modifications (traditional and genetically engineered) of the farmed species and technological developments in harvesting and processing.

Significant competitive 'sorting' among fibre farming countries and regions is likely to occur with these and other developments. In any case, the new-age fibre farms will certainly have the capacity to meet and grossly exceed global needs for bulk wood fibre for the next century, even in face of the rising global population. They can do so, efficiently and almost certainly without any need to use native forests, except as a source for particular specialised and valued wood products that cannot be mass produced under a short rotation agronomic regime.

Consequences of globalisation for stewardship of native forests

So, isn't it a good thing that we can provide for the wood fibre needs of the world without having to manage any of our native forests? Certainly many participants in global and national forest policy debates think so. Substituting fibre farms for cutting in native forests has been a primary goal of environmentalists in Australia for well over a decade. New Zealand took its native forests 'off the table' as a matter of national policy over a decade ago to the joy of most environmentalists.

There are many stakeholders—institutional and individual—that favour 'solving' all our global forest policy debates by substituting exotic plantations for wood production from forests of native species, whether of primary or secondary status. These stakeholders span the spectrum from the wood products industry to hardcore environmentalists and are united in their desire to disengage from the intense conflicts over native forests that characterised the 20th century.

However, a major shift in the global wood products industry from forests and plantations of native species to fibre farming using exotic and bio-engineered species in southern latitudes creates an incredible array of new challenges for society. These include issues of:

- land ownership and use in an environment with decreased potential for economic return from those lands

- stewardship of public forest lands, including provision of the financial resources needed to monitor, protect, and appropriately manage these lands
- the health of the rural communities that are located within forested regions.

For example, in a region like northwestern North America, what will happen to private forest lands, especially the large corporate holdings, when forestry is no longer a profitable enterprise on these lands? Who will acquire them and for what purposes? How will those purposes fit with primary societal needs, such as maintenance of high-quality, well-regulated streamflows or secondary needs, such as open space for recreation and wildlife habitat?

On public lands, where are we going to get the financial resources needed to carry out essential stewardship and how are we going to maintain the skilled woods work forces that will be needed to do this work? How can we retain a capacity to process wood removed from these lands—a capacity that I believe will be important to achieving stewardship objectives on both private and public forest lands. What about the local communities whose fate is strongly linked to that of the forest? And trusts that depend on income from forest lands?

Most temperate forest regions are already struggling with the impacts of massive change, including globalisation, whether or not the ultimate dimensions of the change have been recognised. Some regions, such as the northeastern United States, have already had to cope with the departure of corporate forestry and the disposition of the large tracts of forest that they owned. Societal efforts in such regions to retain forest landscapes and values in the face of such changes provide valuable examples of both the challenges and potential solutions.

The complexities of all of these issues, including some of the potential solutions, are beyond the scope of this essay. However, one issue that I would like to address a little more thoroughly is the need for active management of native forests to maintain native forest function, biodiversity, and health, especially on public lands. As noted above, many stakeholders appear to believe that 'preserving' all native forests resolves our major forestry conflicts and allows us to return their management to 'nature'. I do not agree.

A 'solution' to forestry issues that divides the world's forests into fibre farms and native forests—the former to supply all of our wood products and the latter categorically 'preserved' from active management—I view as potentially dangerous for the temperate native forests of the world. In my view human society will need to be continuously engaged in active management of many native temperate forests even when they are no longer used as a source of wood products. I believe that the proffered 'solution'—fibre farms and preserves—will often lead to undesirable outcomes for native forest function, biodiversity, and health and, consequently, the failure of these forests to fulfill the expectations and needs of human society.

The need for active management of forests

So, why do we need to be concerned about our commitment and capacity to carry out active management of native temperate forests? One very important set of reasons is that we have so altered the physical and biological context from those in which the native forests evolved. A second category of important reasons relates to the societal goals that we are setting for our native forest land-scapes—for example protection of watersheds and maintenance of native biodiversity—that are clearly not likely to be met with *laissez faire* management.

Alterations in the physical and biological context for temperate forests are immense and numerous including:

- altered fire and other disturbance regimes
- altered regional and global environmental regimes, including climatic and chemical changes, such as those associated with acid rain
- introductions of exotic organisms, including virulent insect and disease pests, and other fauna and flora that damage, destroy, and compete with native biota
- fragmented landscapes in which both the amount and spatial pattern of specific forest conditions have been drastically altered.

There are *no* areas of our native temperate forests that are not significantly influenced by these altered physical and biological conditions. And the impacts of altered physical and biological conditions will intensify throughout the 21st century!

Altered fire and disturbance regimes

The issue of disrupted fire regimes provides an outstanding example of the need for human beings to be continuously involved in stewardship of extensive areas of temperate forest. Pre-modern human societies were very effective at modifying natural fire regimes by increasing the number, locales, and timing of ignitions. Modern human societies have been very effective at altering fire regimes in many forest regions by suppressing natural wildfire.

Forest landscapes in western North America provide an example of where natural fire regimes have been dramatically modified during the 20th century. This has been accomplished through a variety of activities including suppression of natural fires, logging of mature and old trees, and active management to create dense ('fully stocked') stands of young trees.

The effects of such activities have been particularly profound in the pine and mixed-conifer forests that evolved under a regime of frequent, low to moderate intensity wildfires.[2]

In these forests, fire suppression has been very effective over the short and mid term. Suppression has resulted in increased stand densities, increased fuel loadings, and greater continuity in fuels (ground-to-crown and crown-to-crown). In many forests, shade-tolerant tree species have largely replaced the shade-tolerant pioneer tree species that are important ecologically and have greater fire resistance. The potential for uncharacteristic stand replacement fires is now very high in many of these forests, putting native forest biodiversity and functions (such as watershed protection) at risk.

Active management of these forests is required to restore and maintain these ecosystems if they are to provide the conditions and services expected by local and global human societies. The negative consequences of allowing wildfire to return to these

---

2 See, for example, Agee, 1993, Sierra Nevada Ecosystem Project, 1996, and Quigley, 1996.

forests without prior treatment, such as by simply suspending fire suppression activities, will often be large and unacceptable. Uncharacteristic stand replacement fires would be a common consequence, resulting in dramatic and uncharacteristic changes in forest composition, structure, and function, including regulation of water and nutrient cycles, and maintenance of habitat and populations of endangered species.

Treatments to significantly reduce forest fuels and protect the resilient older trees are necessary prior to re-introduction of fire to many of these forests Mechanical removal of fuels will often be desirable or necessary. There are also practical and social limits on prescribed fire that will make it impossible to burn all these forests with sufficient frequency to deal with fuel build-ups. There are social issues related to smoke management, and practical issues related to suitable burning periods and availability of trained crews, for example. Hence, significant continuing programs of mechanical treatment, including removal of fuels (logging?), are likely to be necessary in many areas.

Environmental change and exotic pests and pathogens
Detailed cases can be made for the importance of continuing active stewardship in native forests with regards to other stewardship chalenges, such as evironmental change and exotic pests and pathogens. For example, climate changes predicted for the 21st century are going to result in massive geographical shifts in locations of sites that provide optimal and tolerable environments for many temperate forest species; forest communities and types are going to shift locales and even disappear in some. Transitions will often be traumatic, involving stress, accelerated mortality, and even catastrophic loss of existing forests. Relatively rapid climatic shifts and fragmentation of the forest landscapes will make it difficult for migration of native ecosystem components (from fungi to trees) to occur with sufficient rapidity. Opportunities for replacement of native species with undesirable exotic organisms will be enhanced.

Introductions of exotic organisms—from virulent insects and diseases that attack dominant trees to exotic animals and plants that displace native species and alter ecosystem structure and function—represent an extraordinary danger to native forests. Introduction of virulent new forest insect pests and diseases is, in my view, the greatest single threat to the native forests of the world, although regional and global environmental changes will often accentuate problems with pests. This is because exotic pests can so affect a tree species that both its current and future contribution to ecosystem function is effectively 'lost', even when the species is not rendered totally extinct. The effective loss of major tree species is *not* a hypothetical problem! The serious and often catastrophic impacts from introduced pests and pathogens are evident in every temperate forest region of the world. Few native forests have been affected more that those of North America where the litany of exotic pests and pathogens and affected or extinguished tree species is long and getting longer. Consider, for example, white pine blister rust (on all five-needled pines), Dutch elm disease (on American elm), beech bark disease (on American beech), gypsy moth (on many species), *Phytopthera lateralis* root rot (on Port-Orford cedar), and woolly adelgid (on eastern hemlock). Introductions of new pathogens with immense potential to modify native forests appear to be accelerating; the Sudden Oak Death epidemic in California is an extraordinary and as yet not fully assessed example.

Why would any rational society continue to allow the movement of live woody plants, unprocessed wood products (such as logs and wood chips), or untreated wood products (such as green lumber) between continents given this continuing and uncontestable history of introductions? Furthermore, the potential for movement of pests and pathogens between continents presents as great a threat to the fibre farms on which we are going to become dependent for our wood products as it does to the native forests. The exotic tree species that are the basis of fibre farms initially left behind many of their native pathogens but some of the native pests are already 'catching up' with their host species at their new continental locations.

It is imperative that global society halts the intercontinental movement of live plants, unprocessed wood (logs and chips), and wood products that have not been treated (for example, heat dried or chemically treated). Of course, not all the pests are going to be insects and diseases affecting woody plants. An entire treatise could be written about problems created by introductions of other categories of animals (possums in New Zealand, for example) and higher plants and exotic predators (such as scotch broom and gorse in North America and Australia). Suffice to say that such introductions and naturalisations have caused and will continue to cause major modifications of ecosystem composition, structure, and function. These modifications will often be undesirable from the standpoints of maintaining native forest function and biodiversity or of achieving other societal goals.

Perhaps the most relevant point here is that the impacts of changes in environment and introduction of exotic organisms make active management programs imperative. These programs must include substantial efforts and investments in monitoring, research, and resource protection. Furthermore, active management in modifying structure or composition or both is typically going to be necessary to maintain the societal goals for native temperate forest ecosystems.

Fragmented landscapes and degraded ecosystems
Active management of many temperate forest landscapes is also required to restore conditions that will meet societal goals related to goods and services. For example, many of our temperate forest landscapes, including terrestrial and aquatic ecosystems, have been significantly altered by timber management. Native forests have been harvested, dense regeneration stands have been created and managed, extensive networks of roads have been created, and so on. Shifts in management objectives may require restoration of more natural conditions.

For example, federal legislation has shifted management priorities on national forest lands from timber production to conservation of native biodiversity. In northwestern North America, most of the national forests within the range of the northern spotted owl, a flagship old-growth species, have been undergoing significant clearcutting for at least 50 years. Old-growth forests are often highly fragmented. The new management goals for these forested areas call for restoring large contiguous blocks of such forest and sooner rather than later. Hence, significant silvicultural activity is ongoing (and much more is proposed) to treat young forest stands in order to accelerate the development of forest structure characteristic of late successional forest stands.

The diversity and potential scale of management activities to restore terrestrial and aquatic function and structure is very large. While we have much to learn about restoration techniques and their effectiveness, there is no question that active management will sometimes be necessary and will often be very effective, in re-establishing desired structural and functional conditions.

Stewardship of our future forests: recapping the challenge

My goal here is to encourage recognition and discussion of changes in the global wood products industry and the potential impacts of these changes on management of native forests and related human societies. The global wood products industry is moving swiftly towards an agronomic model of economically efficient wood fibre production based on plantations of exotic and, eventually, genetically modified trees grown on short rotations based largely in the southern hemisphere. One consequence will be a drastic reduction in the importance temperate native forests for wood products.

Our reduced dependence on native forests for wood production is viewed by many stakeholders as an opportunity to 'solve' our forestry problems by preserving all or most of the remaining native forests, primary or secondary. Many environmentally oriented stakeholders appear to assume that active management or manipulation of these forests will not be necessary to maintain appropriate functions and biodiversity—that is, that we return these forests to nature and natural processes.

I do not believe that division of the temperate forest estate into fibre farms and reserves is likely to achieve societal goals with regard to the essential goods and services provided by these systems. Active management of many native temperate forests will be essential because of alterations in physical and biological conditions caused primarily by past and current human activities. In effect, passive approaches to management of the many of our native forests—depending on unfettered nature do the job—will lead to unacceptable outcomes from the standpoint of societal goals, including maintenance of native biodiversity.

Forest stakeholders (and most of us are), decision-makers, media, and the general public need to increase their awareness of the fundamental changes that are occurring in the wood products industry and the potential consequences for maintenance of native forest ecosystems, biota, and ecological functions. The evolving societal context, which has the superficial appearance of resolving forest management conflicts, actually creates a whole new series of forest-related societal challenges. These include a clearer definition of our goals for these forests, threats to their integrity in the 21st century, and societal recognition and acceptance of the stewardship responsibilities, including how we will pay for this stewardship.

2

Are forests different as a policy challenge?

Stephen Dovers

Forests have been the most contested environmental issue on political and scientific agendas in Australia and many other countries for many years. Endless inquiries, significant research, blockades, electoral battles, *ad hoc* policy episodes and sustained, intensive lobbying are familiar features rather than occasional occurrences. And the battles seem likely to continue in Australia at least, even after a resource allocation process—the Regional Forest Agreement (RFA) process that was long, intense, and well resourced.

Are forests particularly different and difficult as a policy challenge? Why has contest and acrimony been the norm? Not that these are absent elsewhere, such as in urban development, pollution, greenhouse, fisheries, and so on, but we have made a specialty of division in the forests. To explore that, and to view forest issues through a different lens, I will consider forests as one component of resource and environmental management. Or, rather, as a subsidiary problem in the larger and more profound set of problems that constitute sustainability, known in Australia as ecologically sustainable development (ESD). The specific focus is Australia, but many of the issues are general. A focus on ESD requires a different view than simply ecological or silvicultural management or resource allocation. ESD demands the integration of ecological, social, and economic dimensions, the recognition of cross-sectoral connections, the management of consumption as well as production, taking a proactive or precautionary approach, and the awareness of longer time scales. Not that work on historical, cultural, political, economic, and other dimensions of forests has not been done. On the contrary, the history of division has resulted in a sizeable literature—a research industry even. But comparative analysis of underlying themes relative to other issues is rare.

A selective history of Australian forest policy

Resource management regimes are, like other social institutions, prisoners of history. Just as forest ecosystems evolve in a path-dependent fashion determined by previous conditions, current options for policy change are constrained by past decisions and institutional parameters. With forests, slow cycles of production create even more inertia, as growth and rotation cycles—let alone time frames of ecological processes—transfer the implications of decisions decades into the future. Managing long-lived plants, animals, and ecosystem processes sits uneasily against short-term managerialism. Moreover, crucial institutional elements of forest management regimes—land tenure and property rights—reinforce the inertia, as these are resistant to change. Institutional and ecological time have similarities of length and variability, and contrast sharply with political and economic time.

Major shifts in forest management, and resource management more generally, occur several decades apart at least, with the inter-vening years featuring incremental adjustments to the overall regimes. Australian forest management evidences this trend. Indigenous Australians managed forests deliberately for multiple values, but in ways that white occupiers could not comprehend and that we are only beginning to glimpse. Post-occupation forest management began in the 1790s with colonial regulations to control cedar-cutting. However, early controls and sporadic agitation notwithstanding, the era of modern resource management only began after Federation in 1901 and the appearance of dedicated forest services. Forest management, often in contest for land with agriculture and its attendant government agencies, aimed at ensuring timber supply for national development. Latent concern for other services such as recreation or water catchments had small effect. A pattern emerged of state-owned forest reserves accessed under generous arrangements by private operators. Non-timber uses of forests were catered for under separate tenures, including closed water catchments and a few conservation reserves. Sharpened concerns for assuring timber supplies early in the 20th century, and especially after World War II, led to active participation by state and national governments in plantation development. Forestry on private lands was a small and less regulated part of the sector. This was the pattern of resource ownership, allocation, and management onto which modern environmentalist concerns were projected from the early 1970s. The export woodchip industry also emerged then, involving foreign markets, low prices, unprecedentedly high wood volumes and a more transparent linkage between the public and private sectors. This was a volatile mix and it laid the ground for three decades of bitter contests. These contests were centred around simple definitions of land use and tenure—native forests for wood production in 'state forests' and those for conservation in an increasing reserve estate, with an emerging exotic plantation sector that, many hoped, would ease pressure on native forests. Many incremental adjustments were made, such as concessions toward multiple use and changed silvicultural practices in state forests. Governments did what came naturally, instigating endless inquiries and moderating allocation processes. Some were more substantial, such as Victoria's Land Conservation Council from 1971 and later the national Resource Assessment Commission. These were internationally remarkable institutional responses to emerging challenges that have since been, respectively, weakened and abolished. The national government's use of its control powers over woodchip exports was an important if blunt forest policy instrument. But despite the inquiries, the growing conservation estate, export controls, and forest management adjustments, the timber industry versus conservationist contests did not abate through the 1980s and early 90s, when a new agenda emerged in the shape of ESD.

The ESD process was Australia's response to the global agenda of sustainable development. This agenda forced new considerations across the entire resource management field: protecting ecological life-support systems; the integration of environmental, social, and economic policy; considering longer time frames in decision-making; the need for new policy approaches such as market mechanisms; the precautionary principle; and increased community participation. In the forest sector, the RAC inquiry and the development of a National Forest Policy Statement (NFPS) in the early 1990s co-evolved with the National Strategy for ESD and Intergovernmental Agreement on the Environment. A more principled, inclusive, and sophisticated approach to resource management had been put forward to replace the less constructive contests of the past. During that period another, more powerful policy agenda emerged, informed not by concerns of sustainability but by neo-liberal political ideology and neo-classical economic theory: marketisation. Assuming that public enterprises would be more efficient if they mimicked the characteristics of, or actually became, competitive private firms, marketisation has manifested as privatisation, corporatisation, public-sector downsizing, contracting out, and outsourcing across all policy sectors. Not unrelated is the phenomenon of 'new managerialism', where generic (ie economic) principles take precedence in public-sector management over sector-specific experience or other disciplinary expertise. As a general political and policy trend, and specifically in its best-known form of National Competition Policy, marketisation has driven and continues to drive deep policy and institutional change. It was also, in comparison to ESD, implemented vigorously and well supported institutionally. On the ESD front, there were signs of more cooperative approaches to resource allocation and increased efficiency in environmental protection, and a remarkable increase in community participation through programs such as Landcare. But there were less encouraging signs for the forests, in continued fights, the refusal of key environmental groups to participate in the ESD forest working group, and a logging truck blockade of the national parliament. The single, blunt instrument of export controls had outlived its time, and the RAC–NFPS recommendation for regional planning and allocation was grabbed by governments under siege. The Regional Forest Agreements (RFA) process brings us to the present and stands as the most well-funded and intensive resource allocation process ever undertaken in

Australia. We should now be moving beyond the contests to more cooperative and well-informed policy and management regimes. But fights continue, and RFAs have begun to collapse. Hopes that the RFA process would produce a conflict-free future were misplaced for three reasons. The first reason relates to fundamental characteristics of Australian forests. For a start, we do not have much forest and especially little of high productivity. The demand for use of low relief, high moisture, and nutrient landscapes for timber, indigenous places, farming, urbanism, tourism, and biodiversity far exceed the supply of these services. Old-growth forest, with its special amenity and biodiversity, is very scarce. Humans squabble over scarce things. And people like forests; they are beautiful, rich in life, fun to play in, good for producing water, and highly visible standing or cut down. As a resource to manage, even for few uses, they live a long time and that is difficult. The second reason is that despite is length, breadth, and depth, the RFA process had its faults, arising both from poor process design in the rush of the moment and from the legacy of past decisions. In particular, it suffered from inevitable data shortages; conceptual and methodological challenges in integrating environmental, social and economic dimensions; deficiencies in public participation; lack of attention to private forests and plantations; and tension between 20-year resource allocation guarantees and the need for adaptive approaches in the face of complexity and uncertainty, poorly addressed by (prospectively) unclear monitoring and review provisions.

The third reason is generic and obvious but often overlooked nonetheless. Thinking that one policy episode like the RFA process will achieve closure on any sustainability issue is a misguided quest for instant policy gratification. Sustainability issues as policy problems are characterised by attributes that make them different and difficult: possible ecological limits to human activity; pervasive uncertainty; contested values; complexity; connectivity across problems; poorly defined policy and property rights; lack of accepted policy and institutional options; and multiple spatial and temporal scales. They will not be quickly 'fixed' and require a longer-term, adaptive approach that accepts the experimental nature of policy and management, and the need to monitor, learn, and improve. We are not good at connecting specific policy

processes and events with long-term policy improvement and integration. Those three reasons provide only a partial answer to the
question of what is different about forests, because we have only
considered forests so far. Anything seems unique if viewed in
isolation. We can seek further answers by viewing the forest policy
sector alongside the wider field of natural resource management
and ESD. To do so, I will deal with five themes that emerge from
the attributes of sustainability problems, a history of 'policy
ad hocery and amnesia' in resource and environmental management,
and from other important policy trends such as marketisation and
demands for participatory rather than representative democracy.

Emerging themes

Complexity and uncertainty in time: information, learning,
and persistence
Faced with long time scales, complexity, and pervasive uncertainty,
we should see a high value placed on persistence with policy and
management experiences, monitoring, information, and learning.
While our knowledge base has grown significantly, it is clear that it
is nowhere near sufficient. Overall, Australia's recent, highly
equivocal attitude to R&D, especially that not obviously tied to
short-term economic prospects, is problematic, especially for long-
term public-good issues. In resource management, there are two
related dimensions to this: long-term ecological research and monitoring , and monitoring and evaluation of policy and management
(policy learning).[1] That it has been possible for the RFA process to
produce long-term agreements disrupted in the short term because
we could not get the log volumes right is a worry. However, in
other sectors things are often no better and sometimes worse.
Rainfall stations have been closed in recent years as much because
local post offices have been shut as from any deliberate rationalisation. Stock models in fisheries are problematic still, but then fish
hide under water, swim off when you try to measure them and
often migrate long distances, whereas trees stay put in a most
convenient fashion. We should have a better record of counting
trees than fish. Current policy on dryland salinity is underpinned

---

1   This division and the following arguments are detailed in Dovers 2001.

by inadequate monitoring, data, and predictions, but it is a newer issue and one that, like fish, hides beneath the surface. Soil acidification, recognised as a serious issue in the 1980s and recently rediscovered, was ignored in the interim amid the buffeting of short-term funding and fashions. We do know more about trees than fish and salt, but the overall picture of the forest estate has been deficient for a long time and recent improvements are only partial. Part of that reason for that is the existence of information, but the main issue is sharing and using information between governments, and the record of cooperation in the forest sector is not good. Why is basic monitoring deficient? First, it lacks scientific kudos in terms of publications, external funding and professional standing. Long-term ecological research sites are few, mostly recent, selective and poorly used. Second, it lacks political kudos: it is a long-term, expensive and tiring commitment that does not produce quick media copy. Third, the wave of downsizing, cost-shifting, and reductions in agency capacity at district and regional level has had an impact. The increase in community-based monitoring is marvellous but, when viewed against the decline of traditional agency activities, some cynicism is understandable.

Active monitoring and evaluation of policy and subsequent policy learning might be expected to be commonplace, but it is in fact a poorly understood and disturbingly rare activity. There was certainly a lack of shared learning across the separate RFA processes, and a lack of attempts to learn from a rich history of past regional resource planning endeavours.[2] However, that is no worse than a similar failure—indeed an observable resistance—to using past policy experiences during the compressed design phase of our current headline environmental program, the National Action Plan on Salinity and Water Quality. Policy monitoring and learning are generally deficient in the ESD realm, a result of a lack of human capacity, decreasing staff stability and institutional memory, immediate costs, a lack of methods and routine data capture, and wrong expectations of closure after a new policy. It does not seem that basic environmental knowledge or policy learning in the forests are particularly deficient compared to other sectors. The mistrusts borne of political division and the inefficiencies of a

---

2 For a summary of the great diversity of regional NRM, see Dore et al, 2002.

federal system may appear acute, but it is more likely that they have simply been rendered more obvious. Cooperation between stakeholders and state and national governments over issues such as water allocation or land clearing is no better. To question information adequacy and policy learning so sharply may seem strange, in an age of endless data manipulation, audits, reporting, and the rise of the indicator industry. But it is clear that both these dimensions are deficient in the forest sector, and moreover that they do not connect well. The middle of the information spectrum—summary reports, indicators, and one-off audits—have been embraced at the expense of the two ends: basic environmental data, and the monitoring of policy and management interventions. Moreover, while deficiencies exist in specific resource management sectors such as fisheries, forests, and water, a broader problem for policy monitoring is the fragmented nature of natural resource management, where individuals, agencies, and their experiences are compartmentalised within sectors and issues. Overall, institutional settings do not support an integrated policy and management field, with Australia having relied instead on *ad hoc*, poorly implemented strategies and partial community mobilisation rather than the deeper institutional reform widely regarded as necessary to advance sustainability.

Complexity and uncertainty in space: jurisdictions, tenures, and property rights

Sustainable resource management demands integration across landscapes, jurisdictions, land tenures, and land uses. This is an immense challenge, with agencies, researchers, land owners, and managers confined within specific domains by their training, mandates, legal responsibilities, access to information, and habit.

In the forest sector the difficulties are familiar—managing production versus other uses, managing across landscapes despite state borders, and combining social, environmental and economic considerations. In other sectors, both the avowed intent and obstacles are similar. An 'ecosystem approach' in fisheries, rather than simply managing stocks, is logical, appealing and very difficult. Integrated catchment management has been around for two decades, supported by more institutional reform in Australian jurisdictions than other resource sectors, but is still a far-off ideal.

Integrated management of the coastal zone has been suggested in many inquiries and reports—almost as many as forests have—but is still a work in progress.

The chief barriers to integration of management across the landscape are well known but no less manageable for that familiarity. Ecologically illogical but irremovable political borders, bureaucratic territoriality, divisions among stakeholder groups, incommensurate data sets, poorly developed integrative methods, and a lack of institutional reform to allow whole-of-government approaches. It is not that we are incapable of such whole-of-government approaches or, for example, inter-jurisdictional cooperation, but good examples are exceptions and too often not learned from.[3] A further barrier is a simplistic attitude to tenure and land use somewhat peculiar to Australia: one tenure, one land use. Although we have moved ahead on some fronts—in forests and other domains such as co-management of indigenous and conservation lands—this attitude prevails. Beyond the barbed wire fence is a foreign country to most Australians, and their main access to forests is in conservation reserves rather than production forests, let alone those on private land. The sensible proposition of the High Court that leasehold lands could be managed for both pastoral and indigenous values was not recognised for its historical and legal reality or its inherent efficiency and optimisation as a land-use strategy, let alone its moral force. Rather, it led to a misinformed and acrimonious political debate and to legal and cultural regress. This is not to downplay the legalities and practicalities of multiple use and integrated management, but rather to sharpen understanding of the barriers to it.

Policy and management expertise: still shaping the toolkit

The previous two themes suggest that a difficulty encountered across resource management sectors is a lack of widely accepted policy, institutional and management frameworks, methods and techniques, and this is further emphasised in the following two themes. Debates are ongoing—and largely unproductive—over the

---

3 For example, whole-of-government mechanisms for ESD are belatedly appearing in some states, and Crabb (2002) has identified successful inter-jurisdictional resource management arrangements.

relative merits of broad policy instrument options such as regulation versus eduction versus market mechanisms. The veracity of inference by various methods from incomplete data sets is contested. The scientific validity of competing ecological surrogates and convenient short-hand concepts (keystone or focal species, vegetation associations, etc) are just as contested, and the more interesting questions of their utility in management and the risks attached to management commitments based on them scarcely discussed. Implementing integrated catchment management in a well-articulated and operational fashion remains far easier than just espousing it. Also, tensions between the hydrological determinism of integrated catchment management and many other related policy areas (transport, biodiversity, regional development, forests, infrastructure, etc) are only just being recognised. Methods to assist the core sustainability task of integrating environmental, social and economic dimensions are available (multi-criteria analysis, strategic environmental assessment, non-market valuation, deliberative techniques such as citizens' juries, etc) but not often used and rarely tested comparatively. Adaptive management is an alluring prospect but still unclear and open to abuse. In the early 1990s it was put

> ...[that] though there are no established methodologies either for policy analysis ex post facto or for the evaluation of proposed policies, there does exist a grab-bag of useful perspectives and techniques, analytical and evaluative. The practice of environmental policy studies will undoubtedly, over time, lead to their refinement. Eventually, there may emerge approaches with broad acceptance and proven efficacy.[4]

Since then the situation has advanced, but significant effort is still required to evolve the 'grab-bag' into a widely appreciated toolkit. Unfortunately, individuals and agencies tend to champion one option, through convenience, familiarity or disciplinary bias, rather than appreciate the richness and imperfection of the toolkit and make well-justified choices according to the context and task at hand.

Elements of the grab-bag have been as often applied in the forest context as elsewhere. The deepest yet most unappreciated value of the Resource Assessment Commission, including in the

---

4 Walker, 1992, p253.

forests inquiry, was to apply emerging techniques publicly, and to an extent this occurred in the Regional Forest Agreement process. It is not apparent that a lack of analytical and prescriptive techniques retards forest policy and management any more than other sectors.

Current policy fashion 1: deliberation and participation

In recent years the advocacy of participatory, deliberate, or discursive democracy by theorists and non-government groups has emerged, following the questioning of the effectiveness and equity of traditional representative democratic structures. We have also seen, in Australia more than anywhere, a blossoming of community participation in resource and environmental monitoring, epit omised but not limited to Landcare. The two are not unrelated, but the greatest driver has been community demand for involvement in resource management, and a perception that previous regulatory and educative responses had not achieved enough. The Australian experience has been remarkable and is the world's leading laboratory for community-based approaches.[5] While this experience is enormously encouraging, uneasiness is emerging. The withdrawal of traditional extension services, cost-shifting by governments, little devolved decision-making power, and a lack of support beyond insufficient, short-term funding have all raised suspicions of passing the buck rather than genuine empowerment.

One sector where the increasingly favoured community-based group model has had very little impact is in the forests. Largely, it has applied to private rural lands or to monitoring of common problems such as fauna decline, salinity, or water pollution. Transfer to the historically defined tenure and use patterns of forests is difficult. However, involvement in local management or monitoring groups is only one option for public participation in a liberal democracy. Protest—chained to a tree, writing a letter or blockading Parliament—is one of many forms of public participation in policy. So are voting, membership of interest or lobby groups, submissions to development control or environmental assessment processes, or membership of (or representation by members of) advisory bodies. RFAs featured technical committees involving representative stakeholders and regional forest forums

---

5   A summary review of Landcare is offered by Curtis, 2002.

involving a wider suite of interests. Before that, broad policy devel-opment via the RAC inquiry, ESD working group and National Forest Policy Statement process involved non-government groups. While community-based management groups are less apparent in the forest sector, as in other sectors there has been a move away, especially at the national level, from ongoing dialogue and consensus-building processes involving a broad range of interests in policy development. Some in the forest sector would claim that dialogue is difficult and consensus unlikely given the history of bitter division. That may be so, but early in the RFA days that claim was put during a meeting to consider stakeholder involvement in the process. A seasoned official from the water sector countered the claim by describing the difficulty—and degree of success—in processes to allocate scarce water. An urban planner fresh from a public meeting between industrial developers and local residents might also feel that forests are not so special. One the other hand, the development–conservation division seems clearer in forests, even though jobs versus environment is a frequent division elsewhere. One factor is the strong involvement of the labour movement in alliance with industry, as this is far less apparent in other sectors.

However, there is more than one 'forest industry', even though this may not have been evident in the way policy processes have been structured or the way lobbyists have organised. Native forests versus plantations is an obvious cleavage, as is public versus private sector. So is that between large (for example in the woodchip sector) and small-scale private interests (for example local sawmills), a division which emerged during the RAC inquiry but less obviously since. Another is between wood supply and processing. In other sectors, shifting and strategic alliances across and within the 'industry-conservation' spectrum have proved healthy for policy advance. Examples are the National Farmers' Federation–Australian Conservation Foundation alliances over land degradation, or the occasional commercial–recreational fisher coalitions over coastal development or pollution issues.

Whether forests are particularly difficult or not, structures and processes for community involvement are imperfect in both theory and practice. Two challenges are worthy of note. One is how to maintain and persist with participatory approaches, rather than seeing participation switched on and off when it suits the agenda of governments or when interest groups swap opportunistically between shared discussion and sharp lobbying. Related is the issue of *certainty* in resource allocation, which should be viewed as unlikely in the longer term, as opposed to certainty in the *processes* of negotiation and allocation that we should be able to provide. The second is to carefully select participation strategies in different times and places, taking into account community diversity, different purposes of participation (management, policy formulation, technical advice, monitoring, and so on) and the many forms of participation.

Current policy fashion 2: harnessing the market
The emergence and strength of neo-liberal political and neo-classical economic thought has manifested in Natural Resource Management in a number of ways, some being evident in the forest sector. 'Market' approaches to policy and management fall into two categories: application of specific instruments, variously termed market mechanisms, price or economic instruments; and reform of organisational and institutional settings. 'Market mechanisms' includes taxes, charges, tradeable resource rights, and so on, all aimed at transforming individual or firm behaviour through price signals, and are claimed to achieve environment outcomes efficiently. Market-oriented reform of institutions includes corporatisation, competitive tender, contracting out, and so on, seeking efficiency through using the private sector or private sector-like processes, while minimising public expenditure.

While advocacy of market mechanisms has been strong, implementation has been limited to small, recent markets for salt, emissions, and water. The prospect of widespread carbon trading in future, including the sequestration service of forests, is potentially a major development. Institutional reform has been more pervasive, with public assets and organisations being privatised or separated from direct public sector control. The trend away from dominant public ownership of forest assets towards private ownership or management is just beginning. The most advanced resource sector in terms of marketisation is water, where National Competition Policy reforms have revolutionised water management. The implications for resource management of market-oriented institutional change have scarcely been explored, including the time frames of ecological management versus short-term financial imperatives; maintenance of monitoring programs; community involvement in management when citizens become consumers; cross-sectoral and cross-landscape integration; and the enforcement of residual environmental protection functions.

Again, there is the weakness of a fragmented policy field, where cross-sectoral learning is rare, as between the water sector where resource rights markets are only just being created and fisheries where they have been in operation longer and have been analysed extensively. For the forest sector, there is the opportunity to examine the experience of sectors where marketisation has progressed further. The models will not be transferable, but general insights regarding designing what may be a profound and perhaps irreversible policy and institutional change may be gained, for example regarding design of the statutory framework, performance monitoring and enforcement, or information availability.

Forests are not that different

While the forest sector has its peculiarities, overall it cannot be judged to be less tractable than salinity, climate change, sustainable fisheries management, and other contemporary challenges. As to why the fights have stopped, the stand-out peculiarities are long time frames, the scarcity of the key resource (high productivity, mature forest), and the labour–industry alliance in opposition to conservation interests. Obvious? Unique? Long time frames are difficult but not insoluble, and we are up against them elsewhere, for example in catchment management, infrastructure planning or superannuation policy. And the scarcity situation is not so unusual; it occurs, for example, in water allocation. It may be no comfort, but we have yet to cope with the fact that the standard strategy of modern politics—re-allocative compromise—is limited when the cake is too small to satisfy everyone. We should recognise this, in order to avoid unrealistic expectations of win-win outcomes or even tolerated compromises, and accept that new, highly innovative solutions or very hard choices are needed.

Constructive management of inevitable division will take efforts from all players, government and non-government. In a celebrated work on resource management, politics for the environment was described as 'bounded conflict'. We have the conflict, but not the bounding of it to make it constructive. In the current era of populist and divisive politics in Australia, forests are not so special in that regard. One clear lesson from history and other policy fields is that no single, short-lived process, like an RFA, will deliver resolution, but that the task is ongoing. As to what might deliver resolution in the longer term, options are in circulation: binding expression of ESD principles in statute law and proper whole-of-government mechanisms such as empowered, inclusive commissions, serious monitoring of the environment and its connection to policy design, and mandatory strategic environmental assessment of all government policy, and so on. However, none of these will deliver quickly, because we are addressing structural inconsistencies between human and natural systems. The causes of unsustainability are rooted deep in patters of production, consumption, settlement and governance. Marginal changes, or short-term, sector-by-sector policy fixes are not enough. Management of whole landscapes and cross-sectoral issues for sustainability means doing more and doing better than before. That requires long-term commitments of human, information and financial resources, and deeper reform of the institutional system within which those resources are deployed.

3

Towards ecological forestry in Tasmania

J.E. Hickey and M.J. Brown

There have been many shifts in focus in Tasmanian forestry over the past 50 years. The trajectory has been one of increasing ecological understanding, adaptive management, and recognition of community concerns about forest use against a social and economic backdrop of declining rural populations, increasing environmental awareness, and increasing reliance on the global marketplace. There has been considerable tension in balancing the delivery of the special values and services that the local community expects from forests with the concomitant expectation of a profitable wood production industry.

The island of Tasmania is Australia's most forested state, with 50 per cent of its land area of 6.8 million hectares being forest. Tasmania's native forests have been mapped into 50 communities that include over 900 vascular plant species. The forest fauna includes 131 forest dwelling vertebrate species and an estimated 35 000 terrestrial invertebrate species. It has a smaller complement of some taxa, for example arboreal mammals (five possum and eight bat species), than the forests of eastern mainland Australia, possibly due to Quaternary glaciations which may have caused a migration of species to warmer northern latitudes. Raised sea levels

Table 1 Development of forest management in Tasmania

Decade	Conservation issues	Reserves	Management for wood production	Silviculture	Biodiversity measures	Matrix management
1950s	negligible	Scenic reserves	Rural post-war industrial economy; sawlog driven; some domestic pulp; area rights, API developed	Selective logging no regeneration treatment	Non-existent	Log and leave
1960s	Few and local	Scenic reserves, some National Parks created	Inventory and mapping; pine plantation expansion	CBS developed	Charismatic megafauna, tall trees	Log and regenerate
1970s	Wilderness; pine plantations; woodchips; soil and water	Increased reserves and National Parks	Woodchip exports; sustained yield calculated	CBS routine	Input from amateur naturalists	EIS (informal)
1980s	Forestry and wilderness, soil and water; World Heritage, National Estate listings	Increased forest reservation, World Heritage listings	Woodchip EIS; Forest Practices Code highland eucalypt forests	Partial harvesting ceveloped for dry and Representative reservation commenced	Employment of conservation specialists;	Forest practices Code; streamside reserves
1990s	Biodiversity including invertebrates and non-vascular plants; chemical use; native forest conversion; alternatives to CBS; Cultural intrusion of plantations	CAR reserves; Private land reserves and covenants commenced	IFM; search for global markets; reduced local customer diversity.	Native forest silviculture matched to forest types; Quality standards auditing; Thinning, hardwood plantations and forest conversion	TSPA Priority species, Permanent Forest Estate policy.	Zoning for special values; wildlife habitat strips and clumps; alternatives to CBS R&D
2000-	'Tasmania Together' process; biofuels and CWD	Gap-filling CAR system and special purpose reserves	Sort yards, global markets; regrowth peeling, solid wood composites, regional indicators	Fit for site and purpose silviculture.	EMS in place; regional indicators **Under development:** Threatened community legislation; cessation of native forest conversion	Landscape forestry; biodiversity spines, plantation nodes; **Under development:** variable retention, variable rotations, stewardship zones

API = Air-photo interpretation, CBS = clearfell, burn and sow; EIS = Environmental Impact Statements, CAR = Comprehensive, Adequate and Representative, IFM = Intensive Native Forest Management; TSPA = Threatened Species Protection Act 1995; CWD = Coarse Woody Debris, EMS = Environmental Management Systems during interglacial periods isolated Tasmania and restricted recolonisation from mainland Australia. However, Tasmania's mammal fauna is abundant, at least partly due to its fox-free status (although as many as 20 foxes are thought to have recently established in northern and eastern Tasmania), and species such as the eastern barred bandicoot thrive, although they are endangered in mainland Australia.

Tasmania's forests are highly productive in terms of wood production. Turnover from the forest industry exceeds $1.2 billion per year and wood and paper manufacturing account for 22 per cent of all manufacturing value in Tasmania.

During the 1990s, the pace of change in the way forestry is conducted in Tasmania accelerated. However, the underlying reason for change has remained much the same: a desire either to balance social, economic, and environmental concerns or, more cynically, for one or other protagonist to gain or retain the upper hand. At the beginning of the 1990s, issues involved more representative reserve systems and effects of silvicultural practices, the problems of vegetation clearance, and questions of individual species management. Protagonists continue to address these concerns, but now also consider landscape level forestry issues and alternative silvicultural regimes for lowland wet eucalypt forests.

The major policy issues have centred on land use, and these have not been successfully resolved despite considerable efforts by all parties. The ill-fated Helsham Inquiry—the Commission of Inquiry into the Lemonthyme and Southern Forests in the late 1980s—left both industry and the conservation movement feeling short-changed. This was despite the post-inquiry intervention by the federal government to increase reservation by two orders of magnitude above the amount of area recommended in the majority report. The arguments in that case were directly about the reservation of designated wilderness forest, and its reputed world heritage value. Industry's view was that there was an unnecessary 'lock up' of wood resources, while the environmental movement felt that the government had under-achieved. This rift appeared to have the potential to be resolved through the subsequent involvement of stakeholders in the production of a Forest and Forest Industry Strategy. However, perceived industry-biased outcomes of that strategy resulted in a walkout by the conservation interests from the round table process, the eventual demise of the Green–Labor Tasmanian Parliamentary Accord, and more disputation. The latest attempt at reconciliation of stakeholder views has been the Regional Forest Agreement (RFA) process, again involving federal government intervention. Science has played a large and visible role in each of these attempts to find a common view but, unsurprisingly in an adversarial situation, science has failed to deliver mutually satisfactory outcomes.

The Regional Forest Agreement was established to:

- provide continuing certainty for wood production

- achieve a comprehensive adequate and representative reserve system

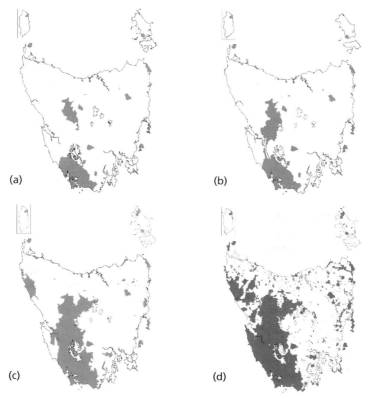

Figure 1 Reservation in Tasmania from 1980 to 2001: 1(a) pre-1981, 1(b) 1982, 1(c) 1992 and 1(d) 2001

- achieve ecologically sustainable forest management both within reserves and in the wood production matrix

- recognise the interests of other stakeholders through a consultative approach with the community.

However, there have been considerable trade-offs for biodiversity maintenance arising from the wood production intensification–reserve allocation model underlying the RFA. Thus about 450 000 hectares of land have been added to the reserve system to bring the total percentage of forests reserved from logging to about 40 per cent of the state's forested land area. The reserves include about 95 per cent of recognised wilderness and 69 per cent of the old growth forests. About 27 per cent of the more productive tall forests are in reserves, the remainder being split between private land (16 per cent) and state forest (57 per cent). On the production side, there has been an increase in intensification of management which will result in the conversion of many thousands of hectares of native forest lands to plantations. The conversion of native forests to plantations, or other land uses, is limited by a Permanent Forest Estate Policy which requires that at least 80 per cent of the native forest estate mapped in 1996 must be retained.

Overall, the RFA process has led to some useful outcomes for biodiversity conservation and forest management, despite the protestations from environmental and industry advocates to the contrary. There has been an increase in the amount and representativeness of forest in reserves, and many of the expert advisory panel recommendations for ecologically sustainable forest management (ESFM) have been put into practice one both production and reserved lands. This is not to say, however, that all the outcomes are optimal. The western third of the island is being managed for conservation largely by benign neglect; it is cold, wet and infertile. The private and urban lands are managed intensively for purposes other than conservation, these lands are warmer, not so wet, and fertile. The third of the island that is state forest, and managed largely for wood production, occupies intermediate sites and is a mosaic of intensive/extensive management for production and some conservation.

Some corollaries of assuming that the current allocation and management achieves good outcomes for biodiversity include the notions that:

- current forest practices maintain biodiversity
- catastrophic and incremental disturbance from global change are catered for
- fragmentation will not lead to species or gene loss
- further vegetation loss won't affect biodiversity
- forests in protected areas do not change over time
- management inside and outside of reserves is complementary
- protected areas are of sufficient size and disposition to maintain all biodiversity.

However, intuition, along with thirty years or so of biodiversity inventory, ecosystem process studies, and silvicultural research all indicate that none of the above hold true. Thus the current management and tenure distribution does not appear to be an ideal method for achieving a harmonious marriage of reservation, wood production, and biodiversity conservation. Decisions about changes to the current land allocation paradigm are matters for society at large to address. The critical agenda is to optimise biodiversity conservation and production within the wood production matrix lands and so move to more ecological forestry.

Silvicultural development

The silviculture of Tasmania's forests prior to the 1950s was characterised by selective logging without any planned regeneration treatment. Regeneration, especially in cut-over wet eucalypt forests, was usually inadequate, except where it resulted from subsequent wildfires. By the 1960s, extensive research into the ecology of the wet eucalypt forests resulted in the systematic application of a silvicultural regime based on clearfelling, slash-burning, and regeneration from seed trees or by artificial sowing. Throughout the 1970s, this regime was extended to most eucalypt forest types with mixed results. Improved ecological knowledge resulted in new silvicultural systems. Partial harvesting techniques were developed and introduced in the 1980s for drier and high-altitude forests with sparse understoreys to encourage retention of useful advance growth and to overcome 'growth check' on high-altitude sites subject to cold air ponding. These forests are frequently multi-aged because of previous selective logging or wildfires. Despite the increased application of partial harvesting techniques where appropriate and feasible, clearfelling with high intensity burning and aerial sowing remains the prescribed silvicultural system for lowland wet eucalypt forests.

Since 1996, there has been a rapid expansion in plantation establishment, mainly through conversion of lowland wet eucalypt forest to plantations of *Eucalyptus nitens* or *E. globulus*. These plantations are managed on short rotations of up to 30 years and require intensive silvicultural practices of site cultivation, fertilising and thinning. About 2.9 per cent of wet eucalypt forests in Tasmania were converted to intensive plantations between 1996 and 2001. In 2000/01, 28 per cent of the harvested native forest was converted to plantations. In the same period, 57 per cent of the harvest was partially logged, while 11 per cent was clearfelled, burnt and sown back to native forest.

Despite the significant amount of plantation conversion of native forest, the public appears to be most concerned about clearfelling, burning, and sowing of wet eucalypt forests, particularly old-growth forests. As elsewhere, the public and scientific debate has been vigorously pursued. The minimum size of opening that defines a clearfell is four to six times average tree height. Where regeneration of wet eucalypt forest is required, clearfelling is followed by a high intensity slash-burn to remove the dense understorey vegetation and logging debris, and to create a seedbed receptive to sown eucalypt seed. Among the perceived advantages of clearfelling and burning are that:

- it fulfills the biological requirements for eucalypt regeneration in wet forests. The system creates a relatively uniform seedbed of heat-sterilised mineral soil over a larger proportion of the harvested area compared with alternative systems.

- wet eucalypt forests, especially *Eucalyptus regnans*, are often even-aged, hence there is often no requirement to retain advance growth

- it has been demonstrated in some forest types to be the safest harvesting system for harvesting contractors, forest workers and machinery

- it requires less knowledge, experience, and supervision to implement than most alternatives

- when followed by slash-burning, it creates a more productive re-growth forest in terms of early establishment and growth of eucalypts. This is related to enhanced nutrition of the seedbed and to higher light intensities on the forest floor.

- slash-burning reduces fuel loads and therefore the risk of wildfire. It also allows more ready access for future forest operations, particularly for thinning.

- it is the most cost-effective method of harvesting and regenerating wet eucalypt forests

- removal of the entire forest canopy ensures free growth of the re-generation.

The perceived disadvantages of clearfelling and burning include:

- creation of an extreme initial visual impact, reducing landscape values and impinging on social amenity

- an assumed disregard for values other than economics and the new eucalypt crop

- effects on biodiversity including at least short-term changes in vascular plant species composition, only sparse regeneration of the rainforest component in mixed forests and loss, or a significant reduction, of old-growth structures and the frequency of those species dependent on older forest

- a reduction in the amount of coarse woody debris and dependent invertebrate fauna

- potential nutrient and organic matter loss from high intensity slash-burns

- higher susceptibility of regeneration to damage and loss of growth or failure due to over-exposure and frost damage on harsh sites

- accumulation and persistence of smoke
- release of carbon into the atmosphere from slash-burns
- loss of advance growth and potential sawlogs in multi-aged forests
- reduction of special species timbers and nectar resources.

There are also concerns that the usual planned rotation length of 90 years is insufficient to redevelop mature mixed forests that have a eucalypt overstorey and a rainforest understorey. Eucalypt forests with a broad-leaved shrub understorey (wet sclerophyll forests) have a notional fire interval of 20–100 years while mixed forests have a notional frequency of 100–400 years. Clearfelling of wet sclerophyll forests at intervals of 80 to 100 years lies within the range of ecological disturbance by wildfires previously experienced by this forest type. However, logging of mixed forests on the same rotation length represents a shortening of the disturbance interval, hence mixed forests may be changed floristically and 'converted' to wet sclerophyll forests. About 20 per cent of Tasmania's wet eucalypt forest is estimated to be mixed forest. Even where wildfires or clearfelling with slash-burning occur at similar intervals, their effect on species composition of regeneration may differ, due to differences in the degree of soil disturbance and compaction, intensity of the fire, the extent and condition of the standing vegetation, and the size of the disturbance.

These findings highlight the need for alternatives to clearfelling and high-intensity burning to be explored. High-intensity burning is currently considered essential for the cost-effective regeneration of wet eucalypt forests and the management of the fire risk posed by unburnt slash. It is also considered an essential precursor to conversion of native forest to plantations. Alternative slash management treatments need to be considered because public opposition to high-intensity burning has increased, based primarily on the nuisance posed by smoke accumulation.

The Tasmanian Regional Forest Agreement noted a priority for research on 'commercial viability of new and alternative techniques especially for harvesting and regenerating wet eucalypt forests and maximising special species timbers production and rainforest regen-

Figure 2 A 10 per cent dispersed retention treatment at the Warra silviculture trial (4 years after harvesting)

eration where appropriate'. The term 'special species timbers' refers to timber from non-eucalypt tree species with the most common being blackwood (*Acacia melanoxylon*), myrtle (*Nothofagus cunninghamii*), leatherwood (*Eucryphia lucida*), celery-top pine (*Phyllocladus aspleniifolius*) and sassafras (*Atherosperma moschatum*).

The Warra silviculture trial was established in 1998 to compare feasible alternative systems with the clearfell, burn and sow (CBS) system, and to develop silvicultural alternatives for areas where habitat, special species timbers, or aesthetic values have additional emphasis. The alternatives being tested include CBS with 3 per cent understorey islands, stripfell/patchfell, 10 per cent dispersed retention, 30 per cent aggregated retention and single tree/small group selection.

The treatments at Warra are being established from 1998 to 2003 and will be followed by a major evaluation, based on environmental, economic, and social performance of all coupes over their first three years, in 2006. However, some preliminary findings are already available. A financial analysis, based on stumpage values, harvest and forest management costs, and modelled growth was carried out for four of the six treatments being trialled at Warra.

These systems were further compared against two silvicultural systems not tested at the site: an intensive regime which includes establishment by CBS, pre-commercial thinning at age 15 years, commercial thinning at age 30 years, and clearfelling at age 65 years; and the option of deferring the harvest of the old-growth forest for a full rotation period. This option is advocated by some special species timber interests as a means of storing these timbers in a living 'woodbank' until they become extremely valuable because of their rarity.

The Willingness to Pay for Land (WPL) concept was used to compare treatments. This is the notional price an investor would pay for an existing forest stand to implement a defined silvicultural system for a designated rotation length over infinite rotations. The results placed the current practice (CBS) as economically superior to all other regimes at the trial. However, based on a 65-year rotation, the regime involving thinning generated higher profit. The analysis did not consider social and environmental values other than to observe that their collective value would need to exceed the difference in WPL between the most economic option (CBS) and some other alternative, for an alternative treatment to be adopted.

Variable retention systems have been developed and implemented in tall wet coniferous forest of the Pacific North West. These commonly involve retaining 20 per cent or more of the forest as aggregates and dispersed trees. The new forest is established by planting into unburnt slash. Variable retention systems may be adaptable for tall wet eucalypt forests in southern Australia but this will be challenging as there are key differences between the two forest types.

First, eucalypt forest canopies transmit more light which allows dense understoreys of shrubs and trees to develop. Mature eucalypt trees have umbrageous or irregularly shaped crowns that are generally unmerchantable. Both sources contribute to huge slash loads after harvesting, sometimes exceeding 500 tonnes per hectare. In contrast, the tall, wet coniferous forests have dense canopies that restrict the development of mid- and understoreys and the trees have spire-like crowns that are merchantable well into the upper crown. Hence, slash loads in the coniferous forests are probably less, although still substantial. Second, the eucalypt forests are generally more flammable, and often occur in more fire-prone envi-

Figure 3 Unburnt slash after clearfelling

ronments than the Pacific North West. It would be extremely poor risk management to leave tracts of highly flammable unburnt slash scattered throughout wet eucalypt forests used for wood production.

However, alternatives to high-intensity burning will need to be investigated for wet eucalypts even though most alternatives attract their own set of environmental, social, and economic concerns. There is potential for fuelwood harvesting (for power generation) to reduce slash loads and allow lower intensity burns and to extend the burning season into mid-autumn. Mechanical disturbance treatments are appropriate for dry forests but become very expensive for wet forests with dense understoreys. Spot or broadcast mulching options require planting rather than sowing, are unlikely to be cost-effective in wet eucalypt forests, and may be less favourable than burning in terms of effects on biodiversity. Planting into dense slash also poses a serious safety issue for tree planters, as well as being arduous and costly.

Variable retention, or other forms of partial-harvesting, are most difficult in old-growth wet forests because of large tree sizes, a high proportion of unmerchantable wood, and dense understoreys. It would be unacceptable if its widespread adoption resulted in an increase in annual fatalities in the forestry sector which currently occur in Tasmania at 1–2 per year. Variable retention systems are more implementable in regrowth forest where they may be more needed. Even clearfelling of old-growth forests usually results in a substantial legacy of diverse propagules and structures (for example stumps, remnant vegetation in gullies, and coarse woody debris) that retain sufficient biodiversity to eventually allow system recovery, at least for vascular plants. However, clearfelling of regrowth, particularly with intensive forest practices of thinning, and shortened rotations to 65 years, will not bestow an adequate legacy for a diverse new forest.

Despite the operational difficulties, there is little doubt that variable retention in wet eucalypt forest will improve biodiversity outcomes in harvested native forest. However, ecologists, environmentalists, certification schemes, and timber markets would need to recognise variable retention as a significant improvement over clearfelling to provide sufficient incentive for forest growers to widely adopt it.

Forest monitoring

Forest monitoring is a key element of adaptive management and continuous improvement. Tasmanian systems for monitoring sustainable forest management have been developed over the last two decades and operate over a range of scales. Most monitoring at the coupe scale is necessarily concerned with inputs—the compliance with prescriptions for wildlife habitat strips and clumps for example— rather than outputs such as the retention of biodiversity achieved one, five or fifty years after a logging operation. The first reporting of indicators for Tasmanian forests was in 2002, based on the period from 1996 to 2001.[1] The report noted the following:

- Thirty-four of Tasmania's 50 forest communities have at least 15 per cent of their pre-1750 extent protected in reserves but 10 communities have less than 7.5 per cent of their pre-1750 extent in reserves. Most of these communities are now primarily found on private land.

---

1   Resource Planning and Development Commission, 2002.

About 0.6 per cent of old-growth forest, mostly tall *Eucalyptus delegatensis* and *E. obliqua* was harvested in the five year period at a rate of about 1000 ha per year.

- Within the wet eucalypt forests, the highest proportions of younger growth stages (< 100 years old) are on private land (54 per cent) and state forest (46 per cent while on conservation tenure 17 per cent is regrowth forest.

- None of Tasmania's forest vertebrates or vascular plants became extinct, but four species presumed extinct were rediscovered during the period. However, one vertebrate (wedge-tailed eagle), six invertebrate species and 20 vascular plant species on the schedule of the Threatened Species Protection Act moved to a higher category of risk. Ten of these were orchids and resulted largely from taxonomic changes. The changed status of the non-orchid species was due to a decline in numbers or increased risks from factors such as habitat loss, *Phytophthora*—a root-rotting fungus, mining, shooting, effect of introduced species, prolonged drought, and inappropriate fire regimes.

While none of these coarse biodiversity indicators is particularly alarming, they give little information on changes at coupe, landscape, and sub-regional scales. Hence their value may be more in making inter-regional comparisons than as an early warning system of significant environmental change. On the other hand, there are dangers in seeking elaborate and costly monitoring systems at too fine a scale, as the results would often have little meaning, at least in the short term. Because budgets are always limited, coupe scale biodiversity monitoring would best be done by a combination of input monitoring and targeted long-term research to investigate areas of potential concern. A Tasmanian initiative on this was the establishment in 1995 of the Warra Long-Term Ecological Research site to investigate the long-term effects of different forest management regimes on natural diversity and ecological processes.

Future directions for forest management

Several projects to assist future planning of the wood production estate are underway. A major one is the Warra silviculture trial to investigate alternatives to clearfelling for wet eucalypt forest. A second project is scoping landscape level variable rotation times to examine the costs-benefits and feasibility of introducing such a system. It is probable that Tasmanian forestry on public land will move in that direction, by using a system similar to the stewardship zoning which is being developed in British Columbia, in which zones are designated with differing degrees of intensification of management to optimise biodiversity and wood production.[2] To some extent, this approach has already been adopted. In southern Tasmania there is a clear gradation of forest management regimes; the eastern portion is private land with towns and farms, and this is bordered by state forest with a mix of plantations on 30-year rotations and thinned native forest regrowth on 65-year rotations. Further west are unthinned native forests on 90-year rotations, and adjoining the World Heritage Area are forests managed on 200-year rotations for special species timbers. Biodiversity is retained, even in the intensively managed landscape, by corridors of unlogged vegetation and by 'biodiversity spines' of native forest that can be logged and regenerated to native forest, but not converted to plantations.

Key biodiversity issues which need to be addressed while acknowledging the need to maintain a legislated sustained yield include the following:

- Adopting a range of silvicultural systems (including clear-felling). Much has been done to deliver variable silvicultures into the drier and upland forests where more than 80 per cent of coupes are now partially harvested. The key focus is now on a greater range of silvicultural systems for wet eucalypt forest. Such approaches have been proposed for wet eucalypt forests but not yet widely implemented.

- Maintenance and recovery of threatened species and communities. There are active studies under way to deal with individual species such as velvet worms and raptors. These projects provide good information on which to base individual management prescriptions which can be implemented under the Forest Practices Code, or adopted at the planning level to counter potential adverse impacts.

---

2 W.J. Beese, G. Dunsworth & J. Perry, 2001.

A more holistic approach is to link landscape ecology with population viability analyses of multiple species having a range of functional life history attributes from across a broad taxonomic spectrum.

- Minimising conversion of native forest to plantations and planning within existing plantation nodes to retain native vegetation either as mini-reserves or as 'biodiversity spines'.

None of the above approaches for threatened species provide directly for the great numbers of known and unknown biodiversity elements that inhabit the forests and for which we have a duty of care to ensure that none become threatened. However, these species can be catered for by changing silvicultural regimes, varying rotation lengths or spatial cutting patterns and/or through planning for the retention and maintenance of surrogates such as adequate supplies of coarse woody debris for habitat.

Forestry in Tasmania has changed markedly since the 1950s when there was little understanding of ecological processes, even of the regeneration processes of the commercial trees. Systems have been developed as ecological knowledge has increased, mainly through research. In the 1960s this was evidenced by the development of the clearfell, burn and sow technique, which has successfully regenerated the eucalypt dominants on thousands of hectares of wet eucalypt forest. In the 1980s ecological knowledge was used to develop silvicultural systems that were better matched to the ecology of highland and dry eucalypt forest. Much of the last decade of research has started to develop understanding of the biodiversity and ecology of the great majority of forest dwelling species that have little or no commercial value, but which are integral to fully functioning native forests. Many of these species are now represented in a greatly enlarged reserve system. However, their existence outside reserves, and sometimes even within them, depends on appropriate management of other tenures, particularly of native forest. It is now time to use the expanded ecological knowledge gained to revise and develop forest management strategies to continue towards ecological forestry.

4

Integrating wildlife conservation and wood production in Victorian montane ash forests

David Lindenmayer

The management of native forests is one of the most socially devise issues in Australia. It has been a major source of conflict and created enormous angst among rural timber communities, government agencies, and conservationists throughout the country. In Victoria, major issues remain over the protection and management of the state's montane ash forests in the Central Highlands of Victoria. These forests are dominated by largely monotypic stands of mountain ash (*Eucalyptus regnans*), alpine ash (*Eucalyptus delegatensis*) and shining gum (*Eucalyptus nitens*). Of these tree species, old-growth stands of mountain ash can include mature and old trees with heights approaching 100 metres, making them the tallest flowering plants in the world. The vast majority of montane ash forests are in public ownership, and the debates surrounding the management of these forests are typical of forest debates around the world.

Montane forests support major timber and pulpwood industries, and have considerable economic value. Several regional industries and local communities are partially dependent on access to timber resources in montane ash forests. The way the forests are cut has major implications not only for biodiversity, but also for water production. These forests also produce most of the water for the city of Melbourne, and timber harvesting within water catchments can affect both water quality and water quantity. There are typically strong relationships between the age of the forest and water yield and water quality. Initially when the forest is cut, water yield is high (because of high runoff) but water quality is low. Yield then declines as large quantities of water are used by vigorously growing young trees. As the forest matures and approaches an old-growth stage, water yields increase significantly and the quality of water is also high. Appropriate consideration of forest management and water management issues could delay the need for a major new dam to support Melbourne's water needs, and could even obviate the need for such infrastructure. Montane ash forests are significant landscapes for biodiversity conservation and, for example, support virtually the entire known distribution of the rare and declining species Leadbeater's possum.[1] In addition, large intact areas of old-growth forest such as those located within the closed water catchments of the Yarra Ranges National Park are critical habitats for charismatic threatened or declining species such as the sooty owl and the yellow-bellied glider. The forests support many species of small mammals and arboreal marsupials, bats, reptiles, and birds, as well as several hundred plant species. Therefore, the intersection of forest biodiversity conservation and wood production is a fundamental part of appropriate forest management. Montane ash forests have been a focus of extensive research in the past two decades for a number of interrelated reasons, including concerns over the conservation of Leadbeater's possum, and controversy stemming from the impacts of clearfelling on environmental values such as the nature conservation. Given the major issues associated with the management of montane ash forests, they provide a useful case study for examining attempts to make a transition toward ecologically sustainable management.

Disturbance regimes in montane ash forests

An understanding of natural disturbance and human disturbance regimes (and how they differ) is part of the key to determining

---

1 Leadbeater's possum is one of the key species of interest in the forest.

Central Highlands

Figure 4 The Central Highlands of Victoria cover about one degree of latitude and longitude near Melbourne in southern Australia

ways to integrate wood production and biodiversity conservation in all forest types, including montane ash forests.

Natural disturbance regimes

Wildfire is the main form of natural disturbance in montane ash forests, and several major fires have occurred in the Central Highlands in the past 200 years, with the largest and most intense one being in 1939. The intensity of wildfires is highly variable. Major conflagrations can be stand-replacing events in which virtually all dominant overstorey ash-type trees are killed. Indeed, large areas of predominantly even-aged montane ash forest presently characterise the landscape. Young seedlings germinate from seed released from the crowns of burnt mature ash-type trees to produce a new even-aged regrowth stand. These processes make it possible to readily determine the age of the dominant overstorey ash-type eucalypt trees in a stand. Although these high-intensity fires can be stand-replacing events, they nevertheless leave many important biological legacies. For example, large-diameter fire-damaged living and dead standing trees occur in many stands of young regrowth, and such trees often contain hollows that provide den and nest sites for many species of arboreal marsupials, as well as bats and birds.

Wildfires also significantly influence conditions on the forest floor. Trees that are killed and collapse onto the forest floor in burned stands become key habitat components for a range of vertebrates. Large decaying logs are also important substrates for the germination of rainforest plants, and the development of dense and luxuriant mats of bryophytes. High-intensity stand-replacing fires represent one disturbance pathway. Lower intensity fires also occur, and these lead to only partial stand replacement because many trees survive. Regeneration of young trees in these forests creates multi-aged stands comprised of ash-type eucalypt trees of two (and sometimes more) distinct age cohorts. Notably, recent studies have shown that up to 30 per cent of stands of montane ash burnt in the 1939 wildfires may have been multi-aged. However, the widespread use of post-fire salvage logging following the 1939 wildfires mean that these stands are now largely even-aged.

Human disturbance regimes

Logging is the predominant form of human disturbance and the traditional harvest method is clearfelling. Virtually all standing trees are removed over 15 to 40 hectares in a single operation. Codes of forest practice allow for up to three adjacent 40-hectare cutovers. Cutting is followed by a high-intensity prescription fire to burn logging debris (eg bark, tree crowns, and branches), creating a nutrient-rich ash seedbed to promote the regeneration of new stands. The rotation time between clearfelling operations is reported to be 80 years, although extensive areas of 45- to 55-year-old ash forest have been harvested in the past 10 years.

Logging and wildlife conservation impacts

One of the major management issues in montane ash forests is the impact of widespread clearfelling operations on biodiversity. Contrasts between the vegetation structure and plant species composition of old-growth stands and young stands recovering after logging have been assessed to forecast some of the impacts of clearfelling in montane ash forests. Statistical analysis of extensive empirical field data shows that old-growth montane ash stands are characterised by:

- numerous large-diameter logs which provide high volumes of coarse woody debris on the forest floor (350 to more than 1000 m³ per ha)
- numerous large living and dead hollow trees that vary considerably in characteristics such as diameter, height and stage of senescence and decay
- an abundance of tree ferns and understorey rainforest trees
- trees of markedly different ages within the same stand, resulting in a multi-aged forest.

In addition, old-growth stands are some of the few places that support features like clumps of mistletoe and associated vertebrate taxa such as the mistletoe bird.

The impacts of clearfelling on stand structure have potentially negative consequences for a variety of taxa. These impacts include:

- significant reduction in abundance of trees with hollows. These trees are nesting and denning sites for arboreal marsupials and other hollow-dependent taxa. Large areas of forest are rendered unsuitable for hollow-dependent animals and the recurrent application of clearfelling on a 50-year rotation ensures these areas will remain unsuitable for the entire suite of hollow-dependent fauna.

- severe depletion of tree fern populations. Tree ferns are important foraging sites for mammals such as the mountain brushtail possum.

- loss of thickets of long-lived fire-resistant understorey plants, including understorey rainforest trees that are nesting sites for birds such as the pink robin

- alteration of landscape composition and the isolation of limited remaining areas of old-growth forest (now reserved from logging) among extensive stands of young forest recovering after harvesting. These changes have negative effects on some wide-ranging vertebrates such as the sooty owl and yellow-bellied glider which are strongly associated with large areas of old-growth forest. Old-growth forests are also important habitat refuges for the mountain brushtail possum and the greater glider, and populations of these species in old-growth stands frag-mented by widespread clearfelling may not be viable in the medium to long term.

In the case of high-profile species such as Leadbeater's possum, field-validated habitat models indicate that the species is typically found in patches of regrowth and old-growth montane ash forest characterised by both numerous large hollow-bearing trees (used as nest sites) and a dense understorey of wattle (*Acacia* spp.) trees, which are a foraging resource for the species. Colonies of Leadbeater's possum are dependent on large trees with hollows that require 200–400 years to develop—a period 5–8 times the length of current clearfelling rotations—and recent monitoring works has shown that populations of Leadbeater's possum are declining (figure 5). This decline is thought to be due to both the natural collapse of existing hollow-bearing trees and the widespread use of clearfelling, which limits the recruitment of new hollow-bearing trees.

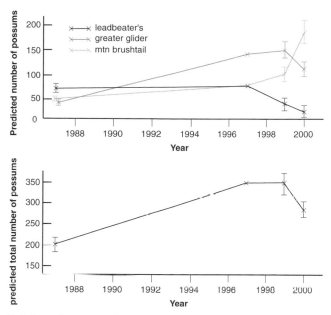

Figure 5 Estimated numbers of Leadbeater's possum, the mountain brushtail possum, and the greater glider (top graph) and the total number of arboreal marsupials (bottom graph) from 1987 to 2000. Non-overlap of 85% confidence intervals (shown) suggests a significant difference between two estimates.

Strategies for biodiversity conservation

Legal and policy frameworks

The Victorian Government has made a commitment to multiple forest use (Government of Victoria, 1986) and, as part of that commitment, the conservation of native wildlife in timber production montane ash forests. The Victorian Flora and Fauna Guarantee Act (1988) is also an attempt to better conserve threatened or endangered species such as Leadbeater's possum as well as other elements of the biota through stating an intention to ensure that '... Victoria's native flora and fauna ... can survive, flourish and retain their potential for evolutionary development in the wild.'[2]

A range of legislative processes can be instigated under the Flora and Fauna Guarantee Act including:

---

2 Victorian Government 1988, p 4.

- adding a given species, community or threatening process to the schedule of the act

- preparing management plans and action statements for the conservation of a species or community or mitigating a threatening process

- invoking an interim conservation order for habitat protection

Notably, the loss of trees with hollows is listed as a threatening process in Victoria, and this is an important consideration given that a wide range of forest-dependent vertebrates depend on trees with hollows; the number of trees with hollows is declining in montane ash forests; and current forestry operations significantly deplete hollow-bearing trees. These factors lead to the undesirable situation that the powers provided in the act have never been used, despite the good intentions of the legislation and the fact that clear-felling is 'legally' a threatening process.

Large ecological reserves

The Yarra Ranges National Park was established in the mid-1990s. It is an extensive reserve encompassing three large water catchments that have been closed to logging for up to 100 years, and containing the most extensive areas of old-growth, mature, and multi-aged montane ash forest remaining in the Victorian Central Highlands. The park supports about 20 per cent of the existing total area of 170,000 hectares of ash-type forest in the region and contributes significantly to the conservation of many elements of montane ash forest biota. However, if it is the only element in a conservation strategy, populations of some species are at risk of extinction from high-intensity wildfires which could affect the entire park. This highlights the need for the conservation of biodiversity in the extensive stands of montane ash forest outside the Yarra Ranges National Park—within wood production areas.

A range of approaches has been taken to conserve biodiversity in wood production areas. There are codes of forestry practice that attempt to limit the impacts of montane ash logging on biodiversity values,[3] and recommendations for the management of high profile

---

3   Department of Conservation, Forests and Lands, 1989; Department of Natural Resources and Environment, 1996.

taxa such as Leadbeater's possum in wood production areas have been made. Other conservation strategies include the establishment of mid-spatial-scale protected areas, wildlife corridors and altered cutting regimes.

Mid-spatial scale protected areas within broadly designated wood production zones

The potential limitations of large ecological reserves make mid-spatial-scale approaches within wood production areas important for the conservation of biodiversity in montane ash forests. Key measures at this scale include the protection of existing stands of old-growth forest, the provision of wildlife corridors, streamside reserves, a management zoning system, and the exclusion from cutting of forests on steep and rocky terrain.

Old-growth forest protection

The protection of areas of old-growth forest is an essential mid-spatial-scale strategy for the long-term conservation of biodiversity in wood production landscapes. The exclusion from logging from old-growth stands has benefits for a range of taxa, and modelling work has indicated that it will benefit a several species of arboreal marsupials. Conversely, a simulated experiment in which old-growth patches were sequentially deleted demonstrated that particular old-growth stands were pivotal for retention; their loss led to a very high predicted risk of the loss of Leadbeater's possum in a given forest block.

The exclusion of logging from old-growth stands was a relatively straightforward task for forest policy makers because the area of such forest is so limited in the vast majority of blocks of wood production forest. Past wildfires (and subsequent salvage harvesting) as well as more recent clearfelling operations have resulted in a predominance of young regrowth stands. Existing 4000- to 10,000-hectare forest blocks devoted to wood production presently contain less than 5 per cent old-growth and are dominated by montane ash stands less than 20 to 60 years old. Few patches of old-growth forest are large—the biggest single patch of old-growth forest (outside the Yarra Ranges National Park) is only 57 hectares. As an example, in the 3500-hectare Murrindindi Forest

Block, the total area of old-growth forest is 24 hectares and it is distributed among eight different patches.

The expansion (restoration) of old-growth stands is important additional conservation strategy in wood production forests because current areas are inadequate. One way to increase the area of old-growth would be to withdraw about 600 hectares of regrowth forest from timber harvesting *within* each forest block and to protect them as multiple reserves that are each 50 to 100 hectares in size. These areas could be allowed to develop into suitable habitat for species such as Leadbeater's possum and the greater glider, although this would be a long process. When restoring the landscape, existing areas of old-growth forest should be the nodal points for future expansion; arboreal marsupials benefit more from regrowth forest reserved adjacent to existing old-growth patches than from regrowth stands set aside far from these areas.

Management zoning within wood production forests

Another mid-spatial-scale conservation strategy that has been implemented within wood production montane ash forests is management zoning. The zoning is designed to assist the conservation of Leadbeater's possum, but it may benefit a range of other species. The approach partitions wood production forests into three types of areas:

- Zone 1: where the conservation of Leadbeater's Possum is a priority
- Zone 2: where wood production is a priority
- Zone 3: where joint land use is a priority

Identification of Zone 1 is based on a broad understanding of the habitat requirements of the species, particularly the abundance of trees with hollows. However, while the zoning system has some advantages, it also has a number of problems:

- It is not based on *all* key habitat components of Leadbeater's possum (such as the abundance of understorey plant species) so that some areas suitable for the species will inevitably be lost.

- The zoning system is temporary. Trees with hollows are lost in montane ash forests as a result of natural collapse and areas of existing Zone 1 forest revert to Zone 2 or 3 and become available for clearfelling when the number of cavity-bearing trees falls below a threshold number.

- The application of clearfelling within former areas of Zone 1 forest on a 50–80 year rotation permanently precludes development of suitable new habitat for Leadbeater's possum; therefore new Zone 1 habitat can not be added in wood production forests.

- On-ground mapping errors have resulted in accidental logging of Zone 1 habitat.

- The zoning system may result in two broad categories of forest in wood production areas—cutover areas aged between zero and 50 years (the rotation time) and old-growth stands greater than 300 years in age. This not only limits the range of age classes in the forest but could limit the recruitment of new areas of old-growth to montane ash landscapes.

The zoning system for Leadbeater's possum has recently been incorporated within a more general forest zoning approach for montane ash forests. High conservation value sites are specified as Special Management Zones (SMZ) and the remainder of the forest where intensive forestry can be practised is called the General Management Zone (GMZ). Despite the renaming of the zones, the five problems listed above still remain.

Provision of wildlife corridors and other set-aside areas

Other areas set aside for conservation (and other) purposes within wood production include wildlife corridors, and terrain that is unloggable because it is too steep or rocky. These areas, together with other exclusions such as old-growth stands and Zone 1 habitat for Leadbeater's possum can comprise up to 30 per cent of a given forest block and they can make a significant contribution to the conservation of biodiversity in these areas. However, like the zoning system and the logging exclusions from old-growth forest, they also have some important conservation limitations.

Some species rarely inhabit narrow wildlife corridors located between cutover areas, although they may use them in moving between reserved areas. The linear shape of corridors is apparently not conducive to their use. Also, excluding logging from up to 30 per cent of the area of individual 4000- to 10,000-hectare timber management units is still inadequate to support long-term viable populations of some taxa. In addition, riparian areas are often dominated by cool temperate rainforest that is unsuitable foraging and nesting habitat for many species.

Stand-level management within broadly designated wood production zones
Retaining trees on logged sites is another important element of an overall conservation strategy for biodiversity in montane ash forests. This tactic helps deal with the potential problems of catastrophic fires in the Yarra Ranges National Park, limitations of wildlife corridors, riparian strips, and zoning, as well as the restricted current extent of old-growth patches. A wide range of species have been recorded in post-logging regrowth forest where numerous large hollow trees have been retained. However, while the approach is valuable, in isolation, it is insufficient for several reasons. Retained trees often are destroyed or badly damaged by high-intensity slash fires, and trees that do remain standing often have poor survival rates. Tree-retention strategies, even if increased by 100 per cent over those presently recommended, will still leave significantly fewer hollow trees in logged areas than occurred in unmanaged stands, and finally, numbers of retained trees also may be insufficient to meet the habitat requirements of a wide range of hollow-dependent taxa.

The need for alternative silvicultural systems

Appropriate forest management practices at multiple spatial scales ranging from large ecological reserves, forest landscapes, and at the stand level are of great importance in the light of the continued widespread application of intensive clearfelling operations. This is because of the significant changes in stand structure which result from the extensive use of this silvicultural system, and the associated impacts on many elements of the biota of montane ash forests.

The results of extensive empirical studies in montane ash forests[4] suggest a need to embrace cutting and regeneration methods in montane ash forests that promote structural complexity in stands of harvested forest to enhance their value for wildlife. Such changes would be entirely consistent with both Victorian and federal government policies and legislation which highlight the need for the integration of wood production and nature conservation in areas designated for timber harvesting.[5]

The Silvicultural Systems Project

Concerns over the impacts of traditional forms of clearfelling in Victoria led to the establishment of the Silvicultural Systems Project in mountain ash and other eucalypt forest types in the 1980s. This was instigated to test and develop economic methods of timber harvesting that also consider other non-commodity values. The broad objectives of the project were: '... to identify and develop silvicultural systems with clear potential as alternatives to the clear felling system and model those systems against clearfelling in terms of the long-term balance between socio-economic and environmental considerations.' A range of types of forest logging treatments were investigated, including: clearfelling, shelterwood, small gap selection, large gap selection, seed tree, and strip cutting. In addition, various methods of seedbed preparation were examined.[6]

Although the project was laudable in exploring alternatives to clearcutting, the study was limited by its use of a restricted set of traditional silvicultural methods. Each treatment resulted in the removal of all stems in a given area on a 50–80 year rotation. For example, one of the shelterwood systems removed retained trees only three years after the regeneration felling. Given removal of all stems, coupled with the requirement for nest sites in large old hollow trees by virtually all species of arboreal marsupials, all silvicultural practices tested have detrimental long-term 'on-site' impacts. It has been established that adequate regeneration can be

---

4 Reviewed in Lindenmayer & Franklin 2002.

5 For example, Government of Victoria 1986; Commonwealth of Australia and the Department of Natural Resources and Environment 1997.

6 For more details, see Squire et al 1987 and Squire 1990.

obtained by the use of cutting regimes other than clearcutting, and high-intensity slash fires.[7] In addition, recent work has clearly demonstrated that mountain ash seedlings can regenerate in gaps as small as 0.1 hectare (compared with the typical cutover sizes of 10–40 ha). This indicates it is certainly possible to implement modified harvesting regimes using a wider range of cutting practices than currently applied. Despite the findings of the Silvicultural Systems Project, alternative harvesting methods have not been widely embraced and extensive clearfelling operations are the norm in all areas of montane ash forest (more than 95 per cent in mountain ash and more than 98 per cent in alpine ash). This means that much of the potential value of the project has been lost because the results have not been implemented. It also means that the current form of logging (clearfelling), which is a key process threatening biodiversity in these forests, has yet to be adequately addressed.

New silvicultural systems

Part of the resolution of the problem of natural resource use conflict in montane ash forests lies in finding more ecologically sensitive ways to cut and regenerate stands. Ecological data gathered from studies of biodiversity within montane ash forests to date indicate that new silvicultural systems should lead to:

- the retention of more living and dead trees that remain standing through several cutting events and eventually develop cavities
- the protection of intact thickets of logging-sensitive under-storey vegetation
- the protection of existing logs, and recruitment of new, large logs to the forest floor
- a shift to longer rotation times more akin with the average natural life cycle of montane ash stands
- the creation of more stands of truly multi-aged forest (an outcome that will naturally arise if prescriptions for the first two points are embraced

---

7 Van der Meer et al 1999.

One approach toward modified silvicultural systems is the 'understorey island' concept of retained vegetation on logged sites that has been developed by workers within the Victorian Department of Natural Resources and Environment.[8] With modifications to include the maintenance of overstorey eucalypts, this could provide the types of habitat conditions critical for many species negatively affected by clearfelling operations. Two advantages of this aggregated retention approach are that reductions in yields from coupes will be minimised if clusters of trees are confined to particular parts of logging coupes, and problems for worker safety during logging operations can be minimised. There are some key problems with the application of a modified 'understorey island with overstorey retention' model. A critical one is the location of retained patches so that damage to them is reduced by regeneration burning of logging slash following harvesting. In many cases, the best place for retained patches will be at the bottom of logging coupes where regeneration fires will be less intense. The movement of logging debris away from the boundaries of retained patches also may better protect them.

Also, the approach will not be suitable for use in all locations; relatively steep terrain, for example, would be inappropriate (because of the difficulties in protecting retained patches from fire). Indeed, the modification of silvicultural systems does not imply a wholesale replacement of one cutting method with another. Rather, it will be important to vary the cutting recipe to suit local conditions. Lessons from natural disturbance regimes can be valuable in this regard. Detailed studies of natural disturbance regimes in montane ash forests have shown that wildfires are likely to be less intense and create complex multi-aged stands (rather than even-aged ones) on flat terrain and deep south facing valley bottoms where incoming radiation levels are low and there is considerable horizon shading.[9] These are areas where serious attempts should made to apply alternative silvicultural systems that lead to within-coupe patch retention.

Notably, multi-aged stands have been shown in studies of vertebrates (such as arboreal marsupials) to be those areas where

8   Ough & Murphy 1998.
9   Lindenmayer et al 1999b; Mackey et al 2002.

the highest levels of species diversity occur. Other investigations have indicated that in the recent past, at least 30 per cent of montane ash forests were multi-aged. This should be the minimum target for stand retention efforts to create multi-aged forests in logged areas. That is, at least 30 per cent of logged areas should be subject to silvicultural systems that result in a multi-aged stand structure. This recommendation is based on hypotheses posed by a number of workers[10] that the impacts of human disturbance on biodiversity should be minimised if they are broadly consistent with natural disturbance regimes. Hence, the prevalence of multi-aged stands created by logging should be broadly congruent with the prevalence of multi-aged stands that occur in 'natural' (unlogged) landscapes. The attractiveness of altered silvicultural systems in montane ash forests is that carefully constructed habitat models (which are based on detailed field data) clearly indicate that timber harvesting that leads to retained patches within cutover areas should provide suitable structural conditions for a range of key vertebrates. Therefore, timber harvesting is in the unique position in Australia of benefiting an endangered species—*if appropriate alternative cutting methods are embraced.*

Models for change from elsewhere
The adoption of new silvicultural systems to evolve harvesting methods away from the widespread use of clearfelling would not be unique to the Central Highlands of Victoria. In western North America and Scandinavia, major changes in logging practices have been made within the last 5–10 years. In parts of these countries, clearfelling is being phased out and replaced with more ecologically sensitive harvesting methods. For example, the Variable Retention Harvest System has superseded clearfelling in large parts of Washington State and Oregon in northwestern USA. Similarly, Weyerhauser Limited has embraced more environmentally sensitive alternatives to clearfelling on its extensive holdings in British Columbia and Alberta in Canada, while still maintaining a viable timber and pulp industries. In Sweden, a nationwide transition away from clearfelling has been made within the past ten years.

---

10 For example Attiwill 1994; Hunter 1994.

The environmental and associated forest conditions in these areas are not dissimilar to those experienced in the Victorian montane ash forests For example, the Douglas fir forests of the Pacific Northwest of North America could be thought of (in many but not all ways) as coniferous analogs of montane ash forests. This indicates that some of the North American lessons and experiences from adopting forest harvesting methods alternative to clearfelling may be useful for parts of Australia.

Testing the effectiveness of new models—a proposed new cutting experiment

The actual effectiveness for biodiversity conservation of the alternative silvicultural models outlined above is not known, and carefully designed studies will be needed to effectiveness of such systems if they are to have any credibility with key stakeholders. In the case of the understorey island with overstorey retention model, a new study (called 'The Cutting Experiment') began in December 2002 to test the response of a range of vertebrate groups to retained patches of different sizes within coupes of logged mountain ash forest. This study has been discussed in some detail with officers of the Department of Sustainability and Environment, and will involve 'piggy-backing' altered cutting regimes on locations that have already been planned for harvesting. Four treatments will be examined so that sites in the planned experiment will either:

- have 3 x 0.5 hectare retained patches (eight coupes)
- have 1 x 1.5 hectare retained patches (eight coupes)
- be subject to traditional clearfelling methods (eight coupes), or
- remain unlogged (eight 'pseudo coupes')

The study will document populations of birds, mammals, and reptiles before logging and then examine responses at regular intervals post logging. All sites will be located in 1939-aged regrowth mountain ash forest—the primary wood production age class and forest type in the Central Highlands of Victoria.

The advantage of the experiment is that through careful identification of suitable sites and by using existing harvesting schedules, the costs of this rigorously designed and powerful study can be minimised. This increases the likelihood that it will continue to be funded over timeframes that are sensible for reasonable forest impact studies. By taking such an approach, it is hoped that the experiment will not suffer the same fate of the very expensive Silvicultural Systems Project which was wound back substantially, thereby limiting the extent of information that could be derived from it. The other advantage of the Cutting Experiment is that it provides a real-world example of alternative cutting methods, driven by forest managers for adoption by forest managers elsewhere in the montane ash forest estate. Hence, the experiment has the potential to act as a type of 'model forest'.

Issues arising from the adoption of new silvicultural systems

There are several important issues associated with the adoption of new silvicultural systems in montane ash forests. The first is that the retention of trees, understorey vegetation, and logs will lead to reduced timber and pulpwood yields from cutover sites. Hence, forest managers may need to revise yield projections to better accommodate other values such as biodiversity conservation. However, the extent of these reductions could be minimised if aggregated (rather than dispersed) vegetation retention patterns are adopted and trees and understorey plants are retained as several consolidated patches within harvest units (as in the understorey island with overstorey retention model described above), and the boundaries of logging coupes are adjusted so that coupe sizes are increased marginally to offset the area of forest set aside within retained patches.

A second important issue is that stand retention practices will add greater complexity to forest planning and operations. However, this should be regarded as a normal part of modern forestry with such complexity being a challenge rather than an impediment to forest and wildlife management. A third issue is the one of worker safety—a critical aspect of any application of altered silvicultural systems. Aggregated retention strategies can help to enhance worker safety (while at the same time improving other values). The increasing use of mechanical harvesters also may promote the worker safety, and small eucalypt patch retention strategies in the radiata pine plantation estate in southern New South Wales, where mechanical harvesting is widely applied, offer some guidance.

Here, eucalypt patches as small as 0.5 hectare can be retained within compartments of plantation pine without jeopardising worker safety. These same patches have considerable conservation value for many (although certainly not all) species.

Other key issues in the montane forests

In addition to the need for modified silvicultural systems in at least 30 per cent of areas of montane ash forest subject to logging, there are other key issues that require attention in the Central Highlands of Victoria.

Biodiversity logging in montane ash forests

There are substantial areas of 1939-aged regrowth montane ash forest throughout the Central Highlands of Victoria. The *Acacia* spp. understorey is dying in many parts of the forest because many of these trees have reached the end of their natural lifespans. There is also a rapid rate of natural collapse of trees with hollows from all areas of 1939-aged regrowth, meaning that the suitability of habitats for wildlife in these areas is being depleted. However, it may be possible to improve the habitat value of some areas of this forest through carefully executed lower-impact logging operations (ie *not* traditionally applied clearfelling), and models such as the biodiversity thinning approaches applied in other regions, such as Douglas fir forests of north-western North America might provide useful guides. One obvious place to start limited trials of such special stand manipulation practices would be within forests in Special Management Zones (see above) that have been assigned particular management status for other values like wildlife conservation.

Additions to the reserve system and the reserve/production forest interface

Part of the montane ash forest estate has been assigned formal reserve status; that is, it occurs in the Yarra Ranges National Park. This is an important part of the strategy to conserve biodiversity in the region. However, some important additional areas need to be included in the reserve system. The Armstrong Creek Catchment is a logical one—if only for the reason of enhanced water production, although there would also be some also significant gains for biodiversity conservation. Including the forest in the Acheron Valley to better link forest between the Maroondah and O'Shannashy water catchments would also add value to the reserve. This presently formally unprotected area has considerable forest on flat or undulating terrain where multi-aged stands (of considerable biodiversity value) are not uncommon.

A key issue associated with the management of the Yarra Ranges National Park is the boundary with surrounding wood production forest. Presently, intensive clearfelling operations take place right to the boundary, and this has the potential to create two types of related problems: an influx of weeds from logged areas (eg blackberry) into the national park, and the risk of regeneration fires escaping from logged sites into the national park. Given these potential problems, it seems appropriate for a buffer of unlogged forest to separate land in the two broad tenures to better protect the integrity of the national park (including its water quality and water production values). There is presently no ecological science to determine what width that such a buffer should be, but it seems likely that buffers of at least the dimensions of streamside reserves (ie 40 metres) would be an appropriate starting point.

Overcutting of montane ash forest

A major issue is the pace of cutting of the 1939-aged regrowth montane ash forest. Regrowth forests have been subject to considerable logging pressure for some decades, and there are real concerns about how rapidly this resource is being harvested. This arises repeatedly during planning the locations of proposed logging coupes under the Wood Utilisation Plans in the region. Indeed, there are always rigorous debates within various arms of the Victorian Department of Sustainability and Environment about the impacts of environmental management strategies on resource availability. The problems facing the department are not entirely of its own making—it is clear that there has been considerable pressure for increased access to forest resources from stakeholder groups such as organised labour (the Construction, Forestry Mining and Energy Union), timber communities (formerly the Forest Protection Society and now Timber Communities Australia), and commercial interests (the Victorian Association of Forest Industries).

The rate of cutting means that stands of montane ash trees are being cut before they reach an age where sawlog (relative to pulpwood) production is maximised. Indeed, most of the wood produced from montane ash forests is pulpwood. This has the potential to create serious distortions in the future products derived from montane ash forests, and it also highlights a need to reconsider rotation times in montane ash forests and make decisions about whether the forest industry will be a sawlog or pulplog-driven one. The potential for significant future overcutting is highlighted by a combination of factors:

- the increasing use of larger (so-called 'B-double') log-trucks to transport logs which results in more wood being carted more quickly
- the trial use of corduroy matting to enable winter logging (and therefore year-round harvesting)
- proposals for recurrent mechanical harvesting of young regrowth stands
- the fact that timber workers are paid by the volume of timber they cut

Such pressures mean that reducing the intensity of harvesting to better meet environmental needs will pose a real decision-making challenge for governments. Notably, in early 2002, the Victorian government announced a 40 per cent cut in the sustained yield of sawlogs in Victoria. However, the montane ash forests of the Central Highlands of Victoria were quarantined from this reduction. It seems likely that the rapid rate of past cutting and the potential for even faster harvesting rates in the future mean that the government will need to make some major and carefully considered decisions on sustained yields in these forests.

Independent auditing of logging operations and environmental management practices

A common problem in the management of many forest types around Australia is that organisations such as those allied with the conservation movement do not trust forest management agencies to adequately enforce codes of practice and other environmental safeguards. Part of the problem lies with the same organisation (here the Department of Sustainability and Environment) being responsible both for timber harvesting and biodiversity conservation—the so-called 'gamekeeper/poacher' syndrome. Indeed, many groups see institutions such as the Department of Sustainability and Environment as simply an extension of the timber industry and incapable of appropriately regulating the activities of private contractors responsible for timber extraction.

Irrespective of the merits of these views, they are nevertheless widely held and more needs to be done to better involve stakeholder groups in auditing how public forests are harvested. One way to do this would be to create site-inspection panels with stakeholder group participants to work with the department to examine the on-ground application of codes of practice and other environmental safeguards. An alternative or additional option may be to divest the department of its twin wood production and wildlife conservation roles and pass the later task to Parks Victoria. This would mean that Parks Victoria has responsibility for nature conservation both within and outside the reserve system, as presently occurs in other Australian states such as NSW.

The need for long-term monitoring and adaptive management

Long-term monitoring is a key part of any attempt to develop forestry practices that are ecologically sustainable. Indeed, it is simply impossible to determine whether conservation strategies and other approaches to maintain environmental values are effective without monitoring. Unfortunately, the record on monitoring is appalling for all government agencies responsible for forest management in Australia (and other parts of the world).

The case in the Central Highlands of Victoria is somewhat different in that Parks Victoria and the Department of Sustainability and Environment have jointly funded a major monitoring program of arboreal marsupials and other vertebrates since 2000; the work was supported by the federal government and private sources before that (and dating back to 1983).[11] Nevertheless, the funding allocation for the current work is made on a largely ad-hoc year-to-year basis which makes forward planning difficult. Moreover, it is

---

11 See Lindenmayer & Incoll 1998; Lindenmayer et al 2002a.

presently unclear whether support for monitoring will be ongoing. This is a critical problem, because without continued support for the monitoring program that is already in place, the Victorian government would be in no position to claim that it is managing montane ash forests in an ecologically sustainable way. This, in turn, would make it untenable to allow the management of these forests to meet certification criteria.

The costs of monitoring forest vertebrates in Victorian montane ash forests has been estimated to be approximately $A70,000 to $A90,000 per year.[12] This is a trivial amount, compared with the value of forest products derived from these same forests. For example, it is roughly 25 per cent of the 1991 estimated value of $250,000 for just a single hectare of wood products from montane ash forests.

An issue closely associated with monitoring is that of adaptive management. Adaptive management involves establishing close links between research and management so that management is constantly improved in response to research findings.[13] For management to be truly adaptive it needs to embrace and implement the outcomes of research. Clearly this has not happened in the Central Highlands of Victoria, where biodiversity and disturbance-related research have highlighted the need for altered silvicultural systems—which have not been adopted. Similarly, many of the valuable outcomes of major government-funded research programs such as the Silvicultural Systems Project remain largely unimplemented.

Rectifying the shortage of naturally-occurring tree hollows
A fundamental and long-term issue in montane ash forests is the shortage of tree hollows—brought about by the natural collapse of existing trees with hollows and the slow recruitment of new hollow-bearing trees (in part restricted by the use of intensive clear-felling operations). This is forecast to have significant effects on a wide range of hollow-dependent fauna in montane ash forests. One approach to tackling this problem is the use of nest boxes as artificial nest sites. This is an expensive solution and seems unlikely to

12 Based on the rotating site selection model adopted by Lindenmayer et al 2002a.
13 Holling 1973.

be effective, given that recent data suggests that nest boxes may not have high rates of usage in montane ash forests.

An alternative approach could be the deliberate damaging of trees to stimulate the development of hollows. This has had considerable success in North America and it may have some merit in Australian forest ecosystems.[14] A number of methods are possible, ranging from tree-girdling to firing decay-inoculated projectiles into trees, but their effectiveness in Australian trees needs to be the subject of detailed research. Nevertheless, in the long term, the perpetual supply of sufficient numbers of trees with hollows will be best met by adopting silvicultural systems that accommodate these requirements.

The status of messmate forests

The focus of most ecological and silvicultural research in the Central Highlands of Victoria has been on montane ash forests. This is not surprising, given the economic value of the resource and the importance of these forests for biodiversity conservation. However, there are substantial areas of messmate (*Eucalyptus obliqua*) forest that occur at the warmer and lower elevation boundaries of the montane ash forests, and these forests are likely to come under increasing logging pressure in the future given the current (and potential future) pace of cutting in montane ash forests. The biota of messmate forests has been very poorly studied and the impacts of intensive cutting in these areas are virtually unknown. Therefore, an important area of future study must be to examine the conservation values of messmate forests and begin the process of developing ways to integrate wood production and nature conservation within them.

The dangers of intensification

A key issue that warrants careful attention in the future, as part of attempts to move toward ecologically sustainable management of montane ash forests, is the need to resist the intensification of harvesting. There have been moves to instigate major industrial forestry projects in native forests in several parts of Australia. These projects aim to use so-called logging waste (eg defective trees and

---

14 For example see Carey & Sanderson 1981; Lindenmayer & Franklin 2002.

logs) for either partial burning to produce charcoal for silicon smelting[15] or complete burning for biomass-based power generation plants.[16] These projects are underpinned by the premise of 'cleaning up' the forest. However, the volumes of wood required to run charcoal factories and biomass power plants will require a significant intensification of logging with a corresponding loss of rotting logs on the forest floor, large old living and dead trees with hollows, and dense thickets of old understorey trees. The resulting simplification of stand structure could, in turn, have major negative impacts on the wide range of species dependent on these key elements of stand structure. On this basis, a critical future debate in montane ash forests must be to head off moves to intensify harvesting. One approach would be to develop minimum prescriptions for volumes of retained trees and fallen logs on cutover sites, and then ensure that prescriptions are appropriately implemented.

Lessons for the future

The montane ash forests of the Central Highlands of Victoria support a well-developed reserve system and a range of important mid-spatial-scale conservation strategies within broadly designated wood production areas (eg wildlife corridors, management zones, and old-growth logging exclusions). The key problem—highlighted by extensive biodiversity and disturbance-related research—is at the coupe level where intensive clearfelling operations are employed. A related problem is that the rate of cutting is substantial and the montane ash resource is over-allocated. This has the potential to significantly curtail future options for management and give the flexibility to on-ground operations (such as altered silvicultural systems) that is needed to better meet nature conservation goals.

While these are some of the best-researched in the world, many of the outcomes of biodiversity and disturbance research have not been adopted at the stand or coupe level. Perhaps the changes have been slow to be embraced because both forest managers and conservationists are quite conservative. Both groups seem to focus

---

15 Environmental Resources Management Australia 2001.

16 J. Raison, personal communication.

their thinking around the land allocation model—reserves for conservation and off-reserve areas for production. However, the integration of conservation and production in areas outside of the reserve system will be critical for sustaining forest biodiversity. The history of research work in montane ash forests makes it clear that alternatives to traditional clearfelling need to be embraced in at least some parts of the forest estate—particularly areas on flat terrain and deep south-facing slopes where wildfires were typically less intensive and created multi-aged stands. Such cutting systems need to result in the retention of patches of trees, understorey, and large logs. Field-validated habitat models suggest that suitable forest structural conditions could be created by logging operations *if* retention harvesting methods are adopted. Therefore, the adoption of altered cutting methods, as well as a long-term commitment to forest monitoring, should be considered as non-negotiable criteria for certification in montane ash forests. The adoption of a cutting experiment with associated monitoring of the outcomes (such as the one proposed above) and which is piggy-backed on existing harvesting schedules provides an economical way to test the effectiveness of alternative silvicultural systems, and also provides a mechanism for true adaptive forest management. The Government of Victoria will need to embrace these sorts of changes because of public pressures that are arising from environmental concerns about montane ash forest management. Indeed, it seems likely that conservation pressures on these forests will increase rather than dissipate in the years to come because of their proximity to Melbourne (and associated tourist traffic) coupled with demands for water from the catchments in the area.

Acknowledgments

It would not have been possible to write this paper without previous collaborative studies in the Victorian forests. I am particularly grateful to a number of leading scientists in this regard including R. Cunningham, C. Donnelly, J. Franklin, B. Mackey, M. McCarthy, A.M. Gill, H. Nix, H. Possingham, R. Lacy, K. Viggers, R. Incoll, C. MacGregor, A. Welsh, and P. Gibbons. The Myer and Poola Foundations are gratefully acknowledged for their support of the Forestry Roundtable Meeting which enabled this paper to be written.

5

The role of large mammals in US forests

Michael Soulé

The structure of many contemporary ecosystems echoes past management decisions. In the United States, it is generally accepted that direct and indirect effects of human agencies have determined forest stand characteristics. Specifically, most ecologists believe that intensive and extensive logging and the habitat-altering effects of human-ignited anthropogenic fire affect the distribution of plant associations, species richness, and the age-structure of forest trees. Moreover, it is widely held that many decades of fire suppression have led to deleterious changes in certain forest types, such as ponderosa pine.

Some social critics, including political ecologists and social deconstructionists, wish to take the argument further and claim that contemporary ecosystems are purely a cultural product because earlier human occupations once altered the distributions and abundances of native species. The flaw in this hypothesis is that virtually all of the species that constitute the flora and fauna of North American forests, excepting exotics, are either found throughout much of the northern hemisphere or endemic to North America; few if any were fundamentally altered by cultural intervention, and none were created by older cultures. In other words, the component species of contemporary ecosystems were neither created nor much affected genetically by the original human settlers. So, when social critics claim that aboriginal peoples 'invented' or 'created' the forest by changing the frequency of fires, by enhancing soils, or by changing the abundance of certain species for economic reasons, it is like saying that rearranging the deck chairs on a cruise ship is the same as building the ship.

The removal of megafauna

Instead of dwelling on the effects of fire and resource extraction, however, I wish to discuss a class of disturbances to forests caused by supernormal levels of plant-eating animals.

Ecologists point out that, globally, overbrowsing by ungulates (native and domesticated) is a significant factor in the structural alteration, simplification, and even the disappearance of forests. North American forests are no exception. The ecological consequences of historical and recent local and regional extinctions of highly interactive species, particularly large predators, from American forests may ultimately be more important than changes in fire regimes. The California poet Robinson Jeffers reflected on the deeper meaning of predators:

> *What but the wolf's tooth whittled so fine*
> *The fleet limbs of the antelope?*
> *What but fear winged the birds, and hunger*
> *Jeweled with such eyes the goshawk's head?*
> *Violence has been the sire of all the world's values.*

This is because the evolution of diversity in plants and animals owes much to predation and the multitude of defensive adaptations that prey have had to evolve. Worldwide, forests are being rapidly depleted of their biological diversity, the large predatory animals being the first animal species to disappear.

The current wave of megafaunal extirpation (local or regional disappearance) is just the most recent such event, however. During the late Pleistocene and particularly in the early Holocene, continental-wide, mass extirpations occurred throughout most of the world, including Australia, North and South America, in much of

Eurasia, and notably in Oceania. I argue here that some of these missing species shaped the evolution of forest plants and maintained forest composition and diversity.

One frequent effect of the loss of predation in forests is an increase in herbivore impacts. Extraordinary levels of herbivory may trigger ecological cascades that can drastically alter the diversity of forests. Therefore, ecosystems deprived of violence in the form of predation typically become degraded and simplified. For example, there is speculation that the disappearance of the North American megafauna, coinciding with the Clovis people's invasion of North America 13,200 years ago, wrought dramatic changes in forests and other ecosystems, changes that we are just beginning to understand.

There are at least three ecological scenarios or pathways that should be distinguished when considering the impacts of megafaunal extirpations or extinctions. One begins with the elimination of most large herbivores; the other two begin with the eradication of most large predators. When herbivores are eliminated, their predators and scavengers are likely to starve to extinction, with subsequent consequences. Note, however, that the reverse does not happen. On the other hand, when large carnivores are selectively extirpated, the immediate response is often the numerical and behavioural release of large herbivores, frequently leading to significant overbrowsing and overgrazing.

A second consequence of control or elimination of large predators has been referred to as mesopredator release. This usually manifests as the behavioural or numerical release of middle-sized predators, which are often persecuted or preyed on by the larger carnivores. Even where mesopredators do not become superabundant, their foraging activities may increase in both time and space, with serious consequences for small prey species such as birds, rodents, reptiles, and insects.

During the last 13,000 years, the megafauna in North America has been subject to four episodes of extirpation and extinction. The earliest event culminated in the virtual elimination of large herbivores beginning about 13,200 years ago, eliminating three-quarters of the mammal genera of large mammals (over 44 kg). This event lasted about 300 years, starting when the Clovis hunters crossed over Beringea to the New World. The Clovis nomads discovered a continent full of large herbivores that lacked experience of human beings and the two-legged, projectile-throwing, indefatigable hunters.

Some have argued that climatic change was a cause of the sudden die off; others believe that diseases brought by the Asian hunters could have played a role. Though the controversy persists, most of the evidence, I believe, favours the human-based explanation. It makes no difference for my argument, however, what the cause or causes were of the die off. What matters is that suddenly large numbers of coevolved, highly interactive ecosystem players disappeared from the ecological stage in North America. This occurred in the blink of an ecologist's eye—from 13,200 years ago to 12,900 years ago. An indirect effect of the herbivore extinctions was the simultaneous disappearance of most large predators, including lions, cheetahs, sabre-toothed tigers, scimitar-toothed cats, dire wolves, and giant, carnivorous birds. In terms of ecological function, the North American ecosystems were suddenly truncated, and the forests were deprived of top-down ecological regulation.

The second episode of megafaunal overkill in the US occurred relatively recently. Beginning in the 17th century, trappers caused a massive decline in beavers and other furbearers because of the market demand for pelts in Europe. The third episode was a transient but profound reduction in populations of large native herbivores during a frenzy of market hunting in the late 19th and early 20th centuries. Herbivores such as the bison, deer, and elk (or wapiti) virtually disappeared from most parts of the US, and the mast-eating passenger pigeon, thought to have been the most abundant bird in the world, met the same fate as the dodo.

At the same time, a campaign to remove large carnivores from the United States was launched by cattle and sheep interests—the fourth episode. It gathered momentum during the market hunting era and is still public policy in many western states. Among the target species have been grizzly bear, wolf, cougar, puma, or mountain lion, coyote, and any other 'varmints' that could be poisoned or shot. One of the indirect results of this campaign is that large areas of eastern deciduous forests in the United States suffer a decrease in species diversity of understory plants, and the failure of recruitment of many species of trees. One of the first things to happen was an explosion of white-tailed deer, whose numbers have increased in the virtual absence of natural predation and in spite of poaching and regulated sport hunting.

These 20th-century episodes of predator extirpation were largely motivated by economics. Livestock grazers wanted to rid the forests and ranges of predators, obviating the need for year-around monitoring and protection of their herds of cattle and sheep. This century-long campaign of trapping, poisoning, and aerial gunning has been subsidised by taxpayer financed hunters and bounties.

This war on predators has been quite successful. Wolves were eliminated from United States, with the exception of two or three states bordering Canada. Similarly, grizzly bears persist in semi-viable numbers only in and around Yellowstone and Glacier national parks. Both species occurred through western North America, including Mexico a century ago. Other carnivores, such as coyotes and cougars, have hung on because they are much more difficult to exterminate than wolves and bears. Predictably, deer and elk herd sizes and numbers increased following the termination of market hunting around 1920. Indeed, the rebound in ungulates has been dramatic.

Consequences of the absence of large predators from North American forests

The effective absence of formerly widespread native carnivores in the United States has led to overbrowsing by native ungulates, particularly highly abundant white-tailed deer, elk, and moose. Abnormal levels of herbivory and easily observable changes in the plant species composition is apparent in many ecosystems. Superabundant white-tailed deer in the Midwest and in the deciduous forests of the eastern US have caused changes in the composition of forests, the local rarity of many herbs and the seedlings of many trees, the transformation of understoreys from mixed herbaceous species to park-like understoreys dominated by near-monocultures of relatively inedible ferns and grasses, and the decline of breeding songbirds. Recruitment of some species of trees and shrubs has been inhibited for decades. These affected species include oaks, white pine, eastern hemlock, northern white cedar, and Canada yew. In the absence of wolves, populations of moose in several localities are now 5 to 7 times higher than those in forests that still harbour wolves, and the wolf-free eastern forests in both the US and Canada are becoming degraded from overbrowsing.

A similar pattern of forest ecosystem simplification is occurring in some national parks in the Rocky Mountain region of the US. The symptoms include gradual loss of some vegetation types, particularly riparian habitat and beaver wetlands. In addition, aspen trees in the northern part of Yellowstone National Park have failed to recruit into the canopy for the last 80 years. What has caused these changes?

In Yellowstone National Park (in Wyoming, Idaho, and Montana) and Rocky Mountain National Park (in Colorado) climate change and the suppression of fires during most of the 20th century have probably contributed to vegetational degradation. But the two most important factors in Yellowstone appear to be the elimination of wolves by around 1925, and the cessation of elk population controls by artificial means after 1968.

Elk have undergone both behavioural and numerical release in Yellowstone and elsewhere. Elk numbers in Yellowstone have grown from about 4500 in 1968 to about 20,000 by 1995, reaching winter densities of 12 per square kilometre in the northern range of Yellowstone National Park by the 1990s. Heavy browsing by elk on willows, a favoured food of beaver, has led to a 60 per cent decline in wetlands dominated by willows since the disappearance of wolves, and beaver and the beaver pond wetland ecosystem have disappeared in the Northern Range of Yellowstone.

Beaver ponds increase the height of water tables, enhancing both willow growth and productivity by increasing inputs of nitrogen and phosphorus. In their absence, runoff is more concentrated, and many streams in the northern parts of Yellowstone are now downcut more than 1 metre below old beaver flowage levels, further lowering water tables. In addition, stream channels are now straighter and less complex. A similar situation occurs in Rocky Mountain National Park where willow patches have declined.

Just south of Yellowstone in Grand Teton National Park, moose populations suddenly increased after wolves and grizzly bears were eliminated, and riparian willow communities were significantly reduced in contrast to outlying regions beyond park borders. Sport hunting of moose is legal in the outlying areas, and this form of predation probably accounts for the difference in browsing intensity. Thus, in the three national parks mentioned, the modification of vegetation and linked changes in animal communities is probably attributable, at least in part, to the eradication of large carnivores.

Other groups of animals appear to be affected as well. For example, 25 per cent of breeding birds surveyed in the Yellowstone region nest mostly in cottonwood, aspen, and willow communities; these plant communities are considered 'hotspots' and 'source areas' for some songbirds. There is speculation that these songbirds might not be viable in the Yellowstone region if these source communities continue to be overbrowsed. Support for this prediction comes from the work of Joel Berger and colleagues who observed a decrease in neotropical migrant bird diversity where there is overbrowsing by moose in riparian willow communities in the Grand Teton National Park. They noted higher bird diversity in control areas just outside of the park as mentioned above.

Biologists speculated that the return of the wolf to these parks would initiate ecological recovery in these landscapes. Wolf reintroduction started in 1995 in Yellowstone, and some models predict that wolves, in combination with other carnivores, will reduce elk populations by about 20–30 per cent below food- and weather-limited carrying capacity, both in Yellowstone and in Rocky Mountain national parks. Wolves now number over 200 in Yellowstone and they, along with other predators, are killing about 2000 elk annually from a herd of about 20,000 in the northern range. Other carnivores are undoubtedly contributing to the limitation of elk numbers although, by themselves, bears, cougars, and coyotes had little noticeable impact on elk numbers before wolf introduction.

The effects of large carnivores on prey behaviour may be more important ecologically than their actual impact on prey numbers. Wolves, for example, can change the distribution of ungulates in time and space (deer, moose) and affect group sizes (musk ox). Although elk numbers in Yellowstone National Park have yet to decline dramatically, some patches of sapling aspen and some patches of willow have been increasing in height since about 1997, suggesting that wolves have reached ecologically effective densities.

The case for restoring large carnivore populations

My impression is that North American ecologists and foresters, when considering ecological base lines or 'normal' amplitudes of ecological change, focus primarily on phenomena such as the variation and periodicity of droughts and fires, as if forests were unaffected by interactions between plants and animals. An obvious exception, however, is insect outbreaks; these are known to cause major, long-lasting defoliation, which, in turn, can lead to tree death, blowdowns, and stand replacement. Though insects such as bark beetles may be the proximal cause of these changes, drought often precipitates such ecological cascades.

Nevertheless, we have no conception of 'natural' baseline conditions or 'normal' variability in forest ecosystems because many of the major ecological players are missing. In fact, the concept of 'historic range of natural variability' for North American forests is suspect because of the instability of species interactions and the absence of large herbivores, including mastodons and giant ground sloths, for thousands of years. Moreover, due to the elimination of predators throughout much of the United States, the harmful simplification of ecosystems during the last two or three centuries appears to have imposed severe constraints on the recruitment of many forest plants. Therefore, the premise that managers can restore forests to normal, coevolved conditions by regulating fire and insects alone is incorrect.

Uncertainty about future ecological dynamics, including interactions between species, is pervasive due to rapid environmental, including climatic, alteration. Attesting to this are the profound changes in small mammal communities that occurred at the close of the last glaciation about 13,000 years ago. We cannot know which interactions and which species will be most critical for the maintenance of biodiversity in the future. In the light of our inability to predict the future, the prudent course is to maintain all native species that ecological and social conditions allow.

These kinds of uncertainty, however, should not discourage managers from establishing operational targets or thresholds for ecosystem recovery. Examples of such qualitative objectives would be:

- the restoration of canopy recruitment of trees and of forest understorey diversity where plant reproduction had been arrested by excessive herbivory

- the recovery of beaver/wetland ecosystems where their disappearance is an indirect consequence of carnivore eradication, and the removal of exotic species that significantly alter the structure, diversity, and dynamics of native forests

Again, the prudent ecological strategy is to maintain all native species. This is because we lack information on the 'prehistoric range of natural variability', so any effort to establish baselines for ecosystem structure and distributions of disturbance is futile. The best management strategy, therefore, is to maintain natural processes such as fire, to preserve and restore all native species— particularly strongly interacting species.

It might behove foresters worldwide to consider the documented impacts of predator removals from North America, Central America, and South America.[1] The simplest proposition regarding forests on other continents subject to past megafaunal extinctions or more recent attempts at predator control is that such forests are today radically different in structure, composition, and species diversity compared to their likely status had large predators (and some large herbivores) not occurred, and if the missing, predator-mediated, ecological interactions were somehow to be restored. In Australia, for example, there is increasing evidence that economically and ecologically harmful changes in forests are being caused by exotic mammalian herbivores, both domesticated (cattle, water buffalo, camels, pigs, goats, and others) and wild (various deer species). In addition, alien mesopredators—the red fox and the domestic cat—have been implicated in the extinction of many small native mammals. In spite of this, there are widespread efforts to reduce the numbers of the continent's largest and most effective mammalian carnivore, the dingo. Such efforts, although they may appear perfectly reasonable to pastoralists concerned about depre-

---

1 See Soulé and Terborgh 1999.

dation, should be questioned or halted until more is known about the ecological effects of the dingo in a variety of ecosystems.

Pastoralists are not the only sector that is hostile to predators, however. Many sport hunters and outfitters are concerned that native predators compete with humans for prey. In addition, many big game hunters claim that sport hunting is a satisfactory surrogate for natural predation, at least on legally hunted species. Even if this assumption were valid, however, its utility is doubtful because the popularity of sport hunting in many industrialised countries, including the US and Australia, is declining steadily, a trend that is likely to accelerate. In the US and Canada, one reason for this acceleration is the rapid spread among deer of chronic wasting disease. Though it is still unclear whether some cases of a similar spongiform encephalopathy in human beings (eg Creutzfeldt-Jacob disease) is caused by consumption of meat from infected deer and elk, it is likely that many hunters will avoid areas with significant rates of infection. Such rates have already reached 13 per cent of mule deer in parts of northern Colorado.

Given (1) the deleterious impacts on forest ecosystems by unchecked populations of large herbivores, (2) the rapidly changing relationship between human beings and large ungulates in the United States, and (3) the probability that that human hunting may no longer be as effective in controlling populations of deer and elk where chronic wasting disease is endemic, foresters should consider advocating the restoration of top carnivores. More aggressive treatments have also been proposed, including the introduction of long-extinct taxa such as elephants to restore ecosystems to Pleistocene-like conditions. Personally, I hope that forestry scientists will become partners in the efforts by conservationists to repatriate the missing predators and mega-herbivores, at least, to North America's ecosystems, for pragmatic if not ethical reasons.

To conclude, violence not only creates species diversity by natural selection, it preserves it by ecological interactions. So one might paraphrase Jeffers:

What but the wolf's tooth shields
The lily and willow, the tree's tender shoots?
What but the cougar's ravenous eye devours the voracious fawn?
And what but fear staves the forest's ruin?
Violence redeems Nature's wealth.

Acknowledgments

I thank Mac Hunter, Carlos Martinez del Rio, and Brian Miller for their suggestions, poetic and otherwise. Much of this chapter is derived from M.E. Soulé, J.A.. Estes, J. Berger, and C. Martinez del Rio, in press, *Ecological effectiveness: conservation goals for interactive species*. Conservation Biology.

6

Resolving forest management issues in British Columbia

Fred Bunnell and Laurie Kremsater

As for many other regions, in British Columbia the 'Earth Summit' or UNCED '92[1] did much to encourage re-evaluation of forest planning and practice. Several international agreements emerged from UNCED '92 that had profound impacts on forestry, among them the *Convention on Biological Diversity*, *Agenda 21*, *Guiding Principles on Forests*[2], and the *Framework Convention on Climate Change*. The agreements reflect public concerns of the time, which were typically expressed as fears about losses. The documents emerging from UNCED 1992 reveal five major fears:

- loss of species
- loss of site productivity (degradation)
- loss of present and future options
- loss of economic opportunities
- loss of local participation and influence in decision-making

---

1 United Nations Conference on Environment and Development (UNCED), June 1992.

2 The term 'guiding principles' reflects the document's complete title: *Non-legally binding authoritative statement of principles for global consensus on the management, conservation, and sustainable development of all types of forests.*

When confronting these fears, concerned environmental groups and the public focus on two primary issues: how much forest (old growth) is protected, and how well forests are managed. One consequence of the complexity and scale of forestry is that non-foresters rarely are able to grasp the role of planning in forestry, so focus their concerns about forest management on specific practices. In addition to these widely applicable fears and issues, British Columbia has specific technical issues that have impeded attempts to address public concern, some of which are shared with Australia. The specific issues have ensured that forestry in BC has attracted global as well as local attention, and suggest a similar future for Australian forestry.

Major policy and technical issues

Changes in policy issues focused on the fears of loss summarised above. Additional technical issues have been created by the history and geography of British Columbia, and six of these and their associates have been particularly influential:

- a relatively short history of logging
- remarkable biological richness
- conditions for sustained conflict
- the critical nature of exports
- few signed treaties with indigenous peoples
- unhealthy labour relations

Brief logging history
There are at least three large consequences of the short period of development in British Columbia's forests: most forest logged is natural, the current flora and fauna appears intact, and policy reflects an abundance of productive forest.

British Columbia's motto became official in 1906—*Spendor sine occasu*. Freely translated it reads 'brilliance without setting', but is more frequently rendered 'splendour without diminishment'. Faith in this lack of diminishment appears to be a provincial trait. Forests in BC were seen as inexhaustible supplies of timber, and the concept of sustainable forestry was not addressed until the Sloan Commission of 1945 stimulated a new Forest Act in 1947. One consequence of a small population and large amounts of forests is that most of the wood harvested within the province still comes from forests that were not previously harvested. Because natural rates of disturbance are often high, these forests may not be old, but they are 'natural' and many have experienced little prior human disturbance. Most of the forest being harvested is in its first rotation, so resources that can decline significantly after long periods of forest practice, such as dead wood, have yet to experience significant declines. As well, large areas of the province remain unharvested. For these reasons, the forest-dwelling flora and fauna appear intact.

The vast amounts of forested land led to policies that currently impede effective forest management. Two of these are particularly noteworthy: the 'social contract' and volume-based tenures. When productive public forest appeared more than sufficient to meet future needs, government policy granted rights to harvest in terms of a 'social contract'. Companies obtained rights to harvest enough public land to encourage construction of mills with the expectation that they would continue to employ workers in good times and bad. That meant the workers remained employed even when companies had to sell wood at a loss. This 'use it or lose it' policy ignored economic realities, did little to encourage intelligent planning, and much to stimulate American arguments that British Columbia was subsidising wood (by selling it for less than it cost to cut it).

Abundance of wood over an area almost intractable to survey well also led to volume-based tenures. With volume-based tenure, as opposed to area-based tenure, wood is harvested up to a given volume *somewhere* (area unspecified) within a larger area. That may work well in relatively homogenous forests from which only fibre is desired. When other values also are sought (eg biological diversity, recreation, visual quality), the actual location of the harvest becomes critical. Typically about 12 to 15 per cent of the area harvested annually on crown lands (94 per cent of the land base) is harvested from area-based tenures. The rest comes from volume-based tenures, thus greatly impeding effective planning for resources other than fibre.

Remarkable biological richness

Other than Nepal, no other north temperate region has the same biological richness. That richness is a consequence of location (straddling three of the five climatic regions for North America), topography (highly mountainous, but with five lowland entry points for adjacent flora and fauna), and prevailing weather systems (perpendicular to mountains). Combined, these three features yield a diverse flora. Of the eight major forest regions described for all of Canada, five occur in British Columbia; eight of the eleven plant hardiness zones described by the US Department of Agriculture for North America occur in BC. In turn, the rich and richly structured flora encourages a rich fauna. Although many species have been severely reduced or extirpated by agriculture and urbanisation, the forest-dwelling floral and fauna of the province appears intact, largely because of the brief history of forest development. British Columbia may be the only north temperate jurisdiction where large predator–prey systems sustain the same species they had 5000 years ago.

As well as being biologically rich, the most productive sites in the province are also part of a vegetation type that has come under global scrutiny—coastal temperate rain forest. Although originally found on almost every continent, only half of the world's temperate coastal rain forests still stand, and the vast majority of that still standing occurs in British Columbia.[3] Trees in these forests live long and grow tall, producing 'cathedral' values. The largest tree species in eastern North American forests once approached 60 metres, and at least 13 species in coastal rain forests surpass 60 metres, some approaching 120 metres. Forests of this stature naturally attract attention, epitomise public fears of loss, and become something 'worth fighting for'.

The richness, including all large carnivores and their prey, creates challenges for sustaining biological diversity. It also introduces more direct challenges to the planning and practice of forestry. At least 15 tree species are commercially harvested in the province. Each has its own requirements for successful reproduction and growth. Moreover, the rugged topography of the province naturally encourages diverse harvesting systems, which in

---

3 See Schoonmaker et al 1997.

turn have to be adapted to the diverse ecological responses of species harvested. One consequence is that effective forest planning and practice must also be diverse.

Conditions for sustained conflict

Three features are necessary for sustained conflict: an educated public, an affluent public, and something worth fighting for. The latter often can be interpreted as appreciable 'old growth', and in the case of British Columbia, it also includes an uncommonly rich and apparently intact forest-dwelling flora and fauna. Each feature is necessary to create sustained, large-scale conflict. Education encourages the recognition of a range of forest values and opportunities that may have been lost elsewhere. Affluence permits the leisure and resources for conflict. European nations have the first two ingredients but have less than 2 per cent of old growth left, have lost their larger carnivores, and show extirpations (local or regional extinctions) or significant declines in many other forest-dwelling species. There is simply not enough older forest left to generate sustained conflict. Conversely, Malaysia still has about 70 per cent of its total area under natural forests, but although relatively well educated by international standards, is only developing the affluence common in the Western world.

British Columbia has all three features necessary for sustained conflict, and we anticipate it will be sustained for decades yet. Moreover, the features stimulating conflict (old growth, temperate rain forest, remarkable diversity) ensure that the conflict has global significance. That in turn means that environmental non-governmental organisations and conservation groups will continue to attract resources to sustain conflict, particularly given the relative affluence of BC's neighbour to the south.

The critical nature of exports

The land area of British Columbia is about 930,000 square kilometres (most of it forested) but the population is still relatively small (less than 4 million). The supplies of wood fibre remain large, and are vastly disproportionate to the needs of the local population. Almost from its inception, the export of wood has been critical to the economy of the province, and the economies of many communities remain primarily resource-based. The amount of wood exported is surprisingly large: for a substantial period, about one-third of global softwood lumber exports tracked by FAO originated in British Columbia.

Given the changes that occurred following UNCED '92, the primary effect of this dependence on exports has been extreme vulnerability to market campaigns. With the provincial economy so dependent on forest products, environmental groups using market campaigns have been able to create unrelenting pressure for change in forest practice. Areas such as the southeastern 'pineries' of the United States also produce large amounts of wood fibre, but most of it is consumed domestically. Both corporations and individuals are much less eager to cause hardship close to home than abroad. Because British Columbia's markets are primarily abroad, and will stay that way, the province will remain particularly susceptible to market campaigns that target foreign markets.

Few signed treaties with indigenous peoples
British Columbia is unique among areas of North America in that very few treaties were ever signed with the indigenous people. The diversity of topography also led to a diversity of First Nations Bands (197 in BC, about one-third of those in all of Canada). The 'Reservations', or land set aside for the use and benefit of a Band, comprises less than 0.4 per cent of the provincial land area. Indigenous peoples are demanding participation in decision-making and a share in resource use encouraged by UNCED '92. One consequence is that the number of outstanding land claims in British Columbia exceeds those in the rest of Canada combined. Recent court cases suggest negotiations will be lengthy, untidy, and confusing, and will change the way forests are collectively managed.

Unhealthy labour relations
British Columbia has the highest portion of unionised workers in North America, and coastal forest workers are among those represented by unions. The historical superabundance of profitable wood (large, old-growth trees) led to a near pathological co-dependency between the unions and companies. For various reasons (including the 'social contract' noted above) companies have acquiesced to union demands, with five major consequences.

First, British Columbia now hosts the highest-paid woods workers in the world. Second, by tacit agreement between unions and companies, higher wages were accommodated by increased mechanisation, thus leading to a decreasing, gradually aging union membership. Third, increased mechanisation has increased capital costs of forestry. Fourth, mechanisation introduced its own environmental problems. For example, moving from high-lead to grapple yarding significantly reduced crew size, but also entailed the widening of roads in steep terrain (because roads became landing sites for the logs). Fifth, the wood so expensively harvested in British Columbia is no longer wanted by much of the world (our 'over-mature' forests in need of rejuvenation have become 'ancient forests' in need of protection; while other areas can produce fibre more cheaply).

Similarities with Australia

Addressing these issues, especially in coastal forests, is not a trivial matter. Not only are they difficult to deal with, but they must be dealt with. From the inception of the province until the present, forestry has been the dominant contributor to the provincial economy, and until recently, about one-third of all softwood lumber exports in the world originated in British Columbia. Because of the global significance of the coastal rainforests, and the development of forest certification (also spawned by UNCED '92), it has been imperative that British Columbia not only practice sustainable forestry, but be perceived to practice sustainable forestry.

These issues manifest themselves differently, or not at all, in different parts of the world. It is somewhat foolhardy to draw similarities between British Columbia and Australia, after only a brief visit to forests of a small portion of Australia. Nonetheless, some similarities in underlying issues appear obvious. The relatively brief history of logging is shared, with similar consequences. Volume-based tenures that impede planning for spatially explicit resources appear common in Australia. Production from natural forests exceeds that of plantations, but is likely to shift in parts of Australia before it does in British Columbia. The current impact of forest harvesting on the native flora and fauna in Australia appears small relative to other regions, and the impacts of lost legacies of natural forests are still largely hidden. There are associated advantages. Spatially explicit planning is much more difficult to implement than volume-based planning, but the gains in sustaining other resources are great and those gains are still accessible in Australia. The flexibility offered by increased production from plantations, and the fact that legacies such as dying and dead wood have not been exhausted, is promising.

The remarkable biological richness also is shared. British Columbia's richness is remarkable within north temperate regions, Australia's richness is simply remarkable. Like British Columbia, much of that richness is forest dwelling and the impacts of forest practice have not yet been profound. There is both time and flexibility, as maturing plantations reduce pressure on native forests. However, a striking and troubling difference exists between the two faunas. Both faunas share a large proportion of cavity-using vertebrates, but in British Columbia there also exists a large group of cavity excavators (such as woodpeckers, flickers, chickadees). In contrast, almost 20 per cent of the vertebrate richness in Australian forests must rely on naturally created cavities, which come only with time and age. It is apparent that without spatially explicit planning for the sustained provision of dead and dying trees, Australia's forest fauna will be gradually depleted.

The two regions also share the three conditions necessary for sustained conflict. The difference to date is that Australia's isolation has ensured some reprieve from serious international conflict. Given the stature of species such as mountain ash (*Eucalyptus regnans*) and the unique species within the Australian forest fauna, that situation cannot be expected to last. The current flexibility fortunately allows opportunity to plan in a calmer fashion before conflict becomes intense. Similarly, we expect relations with indigenous people to become more intense, although probably not as conflict laden as in British Columbia. Still, devising mutually equitable relations with the indigenous peoples is almost certain to complicate forest planning and management.

The remaining two issues noted for British Columbia are not so completely shared, but there are lessons to be learned from British Columbia's experience. Nationally, exports of Australian forest products are not nearly so critical to the economy as they are in

British Columbia, although they may prove to be nearly so for Tasmania, which may come to share British Columbia's current state where market pressures make wood from natural forests difficult to sell. Although the unhealthy labour relations noted for British Columbia seem to be largely absent, there are numerous small communities that rely heavily of forest products and whose aspirations must be part of forest management.

Attempts in British Columbia to resolve the issues

Legislation and planning processes have been enacted to translate public and scientific concerns into improved forest practices, and six of these are elaborated on below.

The treaty process

The British Columbia treaty process is a voluntary process of political negotiations among First Nations (indigenous peoples), Canada, and British Columbia. In treaty negotiations, a First Nation does not have to prove aboriginal rights and title these rights are already recognised and protected by the Canadian Constitution. The main goal of the treaty process is to provide certainty of jurisdiction over land and resources. Through a treaty, the rights and obligations of all parties are set out, thereby resolving conflicting land ownership between the Crown (BC) and aboriginal peoples.

The BC treaty process, open to all BC First Nations, recently had 53 First Nations at 42 sets of negotiations. Most First Nations left the process when the provincial government put its principles to binding public referendum (indigenous people comprise less than 4 per cent of the population and viewed the referendum as 'tyranny by the majority'). These negotiations are arguably the most complex set of negotiations Canada has ever undertaken and the most complex treaty negotiations ever undertaken in the world. The BC Treaty Commission was formed in 1990 to enter the process, and has yet to produce a successful treaty.

Designations of protected areas

British Columbia was the first province in Canada to establish a system of ecological reserves (1971). After UNCED '92, the province elected to expand its Protected Areas System to at least 12 per cent of the land base. In 1991, approximately 6.1 per cent of the land base was dedicated to protected areas; by the end of 1999, approximately 11.4 per cent (10.77 million hectares) was dedicated to protected areas. Since then additional large areas have been set aside as 'option areas'. If only half of these option areas are eventually designated, the total will exceed the 12 per cent target recommended by the Brundtland Commission and endorsed by the United Nations.

The good news is that the Protected Areas System has more than doubled in area, and that considerable consultation was taken with the public, particularly environmental groups, in establishing the additional areas. The bad news is that sustaining biological diversity was rarely a professed goal of those championing particular areas (61 per cent of the protected areas designated since 1991 are subalpine or alpine areas). Moreover, there was little political or collective civil will to establish protected areas in those ecological zones most convivial to human habitation. As a result, BC has an excellent system of wilderness areas that inadequately represent biological diversity. Discussion remains focused on the total amount or area, rather than the distribution of that area. Were the Protected Areas System to embrace 15 per cent of the land base, which is possible, most of the forested land would still be under management. Given the predominant allocation of protected areas to higher elevations, efforts to maintain biological diversity must be focused on the large proportion of the land outside them and employ land use planning and land-use practice guidelines.

Land-use planning

From the inception of commercial forestry in the province, planning of forest operations took place within administrative units that had little relation to ecologically meaningful boundaries. Administrative lines often followed straight lines that crossed watersheds and split broad ecosystem types. Given the enlightenment that came with UNCED '92, it seemed wise to encourage planning within ecologically meaningful units. In 1992, the province embarked on a strategic land use planning initiative. In addition to strategic level planning, a variety of more detailed, planning processes were and still are being undertaken at district or community levels. In all cases, public involvement has been or is being sought. The impetus and the intent were admirable; the outcomes have not been.

Among the problems:

- landscape units were often too small to allow forest planning to incorporate wide-ranging creatures (eg caribou or grizzly bears);

- the landscape planning guidebook was rigorously rule based and, for example, specified keeping certain amounts of each ecosystem in the landscape unit as 'old growth'. These rules applied to each ecosystem type within the landscape unit, further subdividing landscapes into small bits defying effective strategic planning;

- existing administrative units were not *replaced* by landscape units, rather landscape unit boundaries became additional lines on the map. Moreover, other planning processes were superimposed, resulting in a hodge podge of intersecting landscape units, ecosystems, forest districts, company tenure boundaries;

- forest planning had to take place and attain targets within intersections of multiple planning processes that yielded many small units that *no one had planned and which had no rationale*; and

- planning costs increased inordinately.

The Forest Practices Code

Given the policy and technical issues noted above, it was essential that British Columbia develop a world-class code of forest practices.[4] That was undertaken and completed in 1995. It always is easier to focus on practices than on planning, which requires thinking about both time and space. Evaluations of the code concluded that, as written, stand-level practices advocated were superior to most other codes or acts at that time. The Forest Practices Code made reference to planning by referring to a 'higher level plan' that was to be in place. Unfortunately, the higher-level planning guidelines lagged in development and were flawed as noted. Bunnell described the situation in these terms:

---

4 The request that forestry sustain all native species is common despite the fact that urbanisation and agriculture are greater threats to both species and genetic diversity. It probably results because: we will live where we choose, providing food has primacy, and some species survive in forests that were evicted from other ecosystem types.

The biggest failing of the B.C. Forest Practices Code is that a series of guidelines can be applied before the desirable forest structure is determined. That is akin to deciding what kind of house to build on the basis of the sharpness of the carpenter's saw and the strength of his hammer. Usually when we pursue a goal, we select a desired outcome before we consider the tools to aid that outcome.[5]

The Forest Practices Code was the culmination of well-intentioned, rule-based approaches that resulted in guidelines with targets that soon became rigid rules. Legislation was enacted to cover retention of old-growth, protecting buffers around several stream classes, and much more. Maintaining connectivity through using corridors was part of the Biodiversity Guidelines, and was fully implemented on over 500,000 hectares on Vancouver Island alone.

As guidelines became rules, there were four far-reaching consequences. First, the practice of forestry shrank to a 'paint-by-numbers' approach, which denied the professional training of foresters and removed any 'growing edge' from the practice of forestry. Second, the practice of forestry was removed from the context of planning forestry thus making puzzling nonsense of the procedure. These two combined, produced a third consequence. Forcing foresters to 'paint by numbers' and thwarting the ability to plan created an atmosphere of futility that may prove more enduring than any one would like. Fourth, all attention was focused on practices and almost none on outcomes. Compliance monitoring prevailed and effectiveness of the guidelines was largely ignored. Seven years after its implementation, remarkably little is known of the effectiveness of the code—another frustrating legacy of 'painting by numbers'.

Results Based Code

The Forest Practices Code (FPC) did much to improve many practices at the stand level (eg road building, culverts, activities in unstable terrain). It produced unintended and unwanted consequences primarily where these practices had consequences simultaneously enacted though time and space (eg sustaining biological

---

5 Bunnell 1998.

diversity). Such negative effects of the FPC eventually became clear to its creators. As a result, British Columbia has currently drafted a 'results based code' (RBC). One professed goal of the RBC is to move away from the compliance enforced by the FPC, towards management to ensure effective performance around specified values. Were the new code based on outcomes, it would face considerable challenges. While the rules under the FPC were numerous and onerous during planning, they also were administratively convenient and enforceable. Moving to a results- or performance-based code should mean few rules to check, and that monitoring of effectiveness of the forest plan and practices is all important. The few efforts at effectiveness monitoring that have been undertaken quickly revealed how complex sustaining all desired values from forestry is, and the importance of learning from adaptive management (the formal process of refining management through monitoring outcomes of management decisions). Because the government agencies allocated few resources to effectiveness monitoring, and have been much reduced in staff, they have little advice to offer companies who now are responsible for demonstrating effectiveness. Given the issues in British Columbia (noted above), this has the effect of potentially 'feeding the companies to environmental campaigners'.

Under the Forest Practices Code, the government had responsibility for establishing objectives and practices (that they hoped were effective) and companies were simply responsible for following those guidelines. Under the proposed RBC, companies set objectives, implement practices, and are now responsible for demonstrating that those practices are effective at meeting the objectives. They have this responsibility without experience in effectiveness monitoring and few examples to build on where practices have been shown effective. The result will almost certainly ignite more environmental campaigns of the kind that have plagued forestry in British Columbia for decades.

The conundrum forest companies face is that they must demonstrate their effectiveness with little assistance from government researchers and defy the rules-based approach while doing so. That is, the risk is now entirely the companies', for the code does not consider due diligence. It is not enough for a company to have undertaken reasonable attempts at determining best practice if that practice turns out to be ineffective. The conundrum is worsened by the fact that the default rules ('best practices') retained from the former FPC were rarely evaluated for effectiveness. Companies' efforts to learn through adaptive management are themselves hindered by the proposed RBC, whose default position remains decidedly rule based. If companies follow the rules and these prove ineffective, the fault apparently is theirs. Moreover, if the RBC is enacted, it is likely that environmental campaigners will again focus on practices, and the actual planning of desired outcomes will continue its stagnation.

Certification

To sustain market share, forest companies have sought forest certification that will provide international credibility. Three codes of certification dominate within the province: Canadian Standards Association (CSA), Forest Stewardship Council (FSC), and the Sustainable Forestry Initiative (SFI). Many companies have successfully achieved CSA certification, but FSC certification efforts have been slowed by delays in drafting regional standards. Regional standards include indicators relating to First Nations, and these have been slow to evolve among the many First Nations in BC. As well, it has so far proven impossible to define 'high conservation value forests' or appropriate management practices for them. In part, this difficulty stems from the diversity present in BC that makes cultural or biological nonsense of 'broad-brush' approaches. Without regional standards, the interpretation of FSC guidelines has depended largely on the individuals involved in the assessment team. Companies that export sizeable portions of their product to the United States have sought and received SFI certification, which makes few concessions to sustaining biological diversity.

Changes in forest planning and practice

Forest planning and practice have changed rapidly in response to changes in products and values desired from forests. The increase in desired values has led to over 80 resource map overlays in the development of a Management and Working Plan for a Tree Farm License. Although many individuals still embrace the concept of multiple use, there is growing realisation that a given piece of land, unless very large, can not provide all desired, sometimes competing resources. Instead of multiple use, land mangers are increasingly exploring zoning of the land base as a workable approach. The perception is growing that all parts of the ecosystem are irrevocably linked and activities that affect one aspect of the system will likely affect others. The current focus is on managing ecosystems, sustaining biological diversity, and altering forests less from their historical condition. Unfortunately the terms 'ecosystems management', 'biodiversity', and 'natural disturbance' have so many different meanings, that they have lost clarity. We believe it is more helpful to clarify public desires from forests by examining concerns about losses noted earlier in this chapter. Most of the losses noted (species, productivity, future options, and economic opportunities) are integral parts of the concept cluster termed 'biological diversity'. Sustaining biological diversity has become a fundamental goal of forest management.

Scientists and managers have translated public concerns and their own improved understanding of forest systems into new approaches. In the Pacific Northwest, these include 'green tree retention', 'patch retention', and 'variable retention'. Social concerns and the long time periods and large areas are incorporated into the concept of 'ecosystem management'. Concerns about losing species and productivity have impelled policy makers to create legislation (eg British Columbia's Forest Practices Code), integrate recent scientific knowledge,[6] and initiate new approaches to planning (eg Innovative Forestry Practices Agreements). Market campaigns directed by environmental groups with a particular focus on Europe were successful in convincing some companies to adopt retention harvesting systems. Variable retention was prominent among recommendations of the Scientific Panel for Sustainable Forest Practices in Clayoquot Sound (CSP 1995) and was first widely applied by MacMillan Bloedel (now Weyerhaeuser BC Coast) within their Forest Project. Other coastal companies

---

6 See, for example [CSP] Scientific Panel for Sustainable Forest Practices in Clayoquot Sound, 1995, Sustainable ecosystem management in Clayoquot Sound. Planning and Practices. Victoria BC.

have since adopted the approach. In the southern interior of the province, both the topography and ecological responses of the tree species harvested had already encouraged uneven-aged management, which approximates outcomes intended by variable retention. Much of the northern part of the province is sufficiently rural and ignored by environmental campaigners that fewer changes in practice have occurred.

Changes in legislation in British Columbia discouraged clearcutting and promoted a range of retention during even and uneven-aged management. Variable density thinning has been used to hasten the creation of old-growth structural attributes. Many forest plans now rely on a coarse filter approach to accommodate needs of most species and to maintain ecosystem processes, and on a fine filter approach of practices designed to meet needs of a few well-known, forest-dwelling species believed to be particularly vulnerable to forest practices. Coarse filter approaches include targets for amounts of older forests and specific habitat components (eg large dead trees, logs), while fine filter approaches are designed for specific individual species. Given the number of forest-dwelling species, it is fortunate that the forest attributes used in coarse filter approaches link to both public concerns and to species richness.

Whereas stand-level practices are commonly well implemented, planning of larger units—watersheds or broad landscapes—is poorly developed. To a considerable extent this results because both certifiers and the general public focus on practices, which are more readily comprehended than planning. Moreover, the province has chosen to initiate novel planning systems at the same time it is initiating a new Results Based Code. Projecting impacts of stand-level treatments and cumulative impacts of stand management remains in the realm of computer projection models, and the compliance-based history has provided little incentive to examine underlying issues (eg patch sizes, targets for older forests, distribution of dead trees and other forest structures, and connectivity). Conversely, riparian areas have received considerable attention, particularly on the coast where small streams are abundant and their roles poorly understood .

We're not there yet

Despite rather massive changes in forest planning and practice, some of the largest issues remain unresolved, and in some instances represent systemic issues that will remain for decades yet. It is impossible to assign a priority of effect on these, but in all cases effects will be large. Five large issues are noted here. Two appear unique to British Columbia, one appears particularly well expressed in the province, and the other two are probably generic.

The unresolved and unique issues are those relating to indigenous peoples and conditions for conflict. Overarching all landscape planning and most forest management plans in British Columbia are unresolved issues surrounding aboriginal rights and title. Processes are underway to settle land claims, but results will necessarily complicate planning and business relationships. Several joint ventures between forest companies and First Nations have been implemented in the province, and some may provide good examples (eg Iisaak Joint Venture with Weyerhaeuser in Clayoquot Sound). Likewise, all the ingredients for sustained conflict remain, and may have been worsened.

As noted, the proposed Results Based Code legislation appears designed to stimulate conflict, with the only benefit being to the government, which will have removed itself from both responsibility and the fray. Both the requisites and conflict-laden issues remain—particularly the fact that British Columbia will continue to harvest 'old growth' and First Nations rights and titles remain unresolved. The provincial economy will remain vulnerable to perceptions about the conflict, because most wood products will continue to be exported. The new federal endangered species legislation (Species At Risk Act) will add to the conflict and confuse large-scale planning still further. Previous attempts to pass the act have satisfied no one, and will thus prove a magnet for confrontation. British Columbia will receive most of the attention because it is the biologically richest province in Canada and has the largest areas of 'cathedral value' old growth. Few individuals are concerned about inland old growth where trees rarely attain 30 centimetres in diameter.

The issue that appears unhelpfully well expressed in the province is the lack of understanding of forestry evident among those creating legislation. Forestry is the most complex of land use practices, and the public has broadened the demands made of forest systems. Although difficult, adaptive management is possible in forestry.[7] Some practitioners in British Columbia have embraced adaptive management and attempted to learn by doing and monitoring. What has not changed, and may be a systemic problem, is an apparent lack of humility among those creating legislation. The notion that we already know enough to manage for the increased values desired from forests remains explicit as 'best practices', and impedes further learning. Admittedly, there is an alternative explanation to simple hubris. The absence of 'due diligence' from the code may represent recognition that the creators do not really know what might be better practice. It does, however, place the government in the strange position of having little responsibility for what happens on forested land, despite the fact that 94 per cent per cent of forest land is publicly owned.

Two troubling issues within the province may be generic—inadequate attention to planning, and the challenges of coping with the rate of change. Provincial initiatives have once again uncoupled planning and practice, which exaggerates the focus on individual practices and has the effect of attempting to meet all desired values in small areas (which is impossible). When forest planning can be undertaking thoughtfully, many troubling issues can be resolved. For example, when confronting the issue of sustaining biological diversity, a decent ecological classification system and commitment to sustaining a portion of all ecosystem types in an unharvested state will sustain many species. That involves digital ecological mapping and spatial strategic planning over an entire forest tenure, or larger area. Planning for the sustained provision of dying and dead wood can be accomplished at the tactical level, over areas of about 50,000 to 100,000 hectares. Such a hierarchical approach simplifies planning at the operational level and avoids many of the conflicts that arise from focus on single sites out of context of the overall forest structure.

Responding to the challenges or current rates of change is clearly a generic issue. Because many tree species grow slowly, rates of change within a forest appear ponderous on a human scale. The

---

7   See Bunnell, Dunsworth, Huggard & Kremsater 2003.

enthusiasm with which policy makers and practitioners have attempted to meet changes in public desires from forests has been much faster. As a result, the practice of forestry has outpaced forestry education, research, and experience as in no other time in history. That condition is generic. It implies that the legislation and practice of forestry requires a new humility. That has not yet happened in much of British Columbia.

W(h)ither the future

Because of the issues noted at the outset, British Columbia finds itself in the awkward position of producing wood than nobody may want because of associated market campaigns. Moreover, many other regions can produce fibre more cheaply and quickly if markets want it. Except for a tiny area of hybrid poplar, less than 3 per cent of the province has conditions that will allow commercial rotations less than 60 years. Nor will global attention on British Columbia allow more growth of potentially faster-growing exotic species, as is common elsewhere. Nonetheless, the provincial economy remains firmly dependent on the export of forest products.

In short, some aspects of forestry in British Columbia are doomed to wither. The total cut will decline, mills will close, some timber-based communities will be thrust into financial trouble, and the provincial domestic product will decline until alternative revenue sources are developed. Despite that seeming gloom, the province is at the forefront of some issues that will become increasingly common in the practice of forestry. It is not special foresight, but the same broad issues that create problems for forestry that have led the province to the fore in some areas (specifically, the economic reliance on forestry, the richness of the resource, and the enormous pressure for improvement).

There are at least five broad aspects of changing forestry where British Columbia has much to contribute to collective knowledge and experience: 'conservation forestry', zoning the intensity of forest practice, simulation modelling, adaptive management, and effectiveness monitoring. Both internal and external pressures have encouraged development of 'conservation forestry', initially to sustain specific forest-dwelling species, but currently to sustain all desired forest values. Research on hastening the development of old-growth attributes through specific silviculture began more than 20 years ago, and the operational steps were implemented in 1984. In its recommendations regarding the use of variable retention, the Clayoquot Scientific Panel noted more than 40 operational instances of small-scale conservation forestry on the coast of British Columbia, many more were proceeding in the province's interior. Since then the experience has grown, become applied over much larger areas, and been documented.[8] If the struggle for market share involves demonstration of abilities to sustain all forest values, some BC companies are well positioned. It remains unclear, however, whether or not the market will reward such effort. The application of conservation forestry bears economic costs, and works best when employed within a framework that zones the intensity of forest practice or fibre removal.

Many of the early theoretical arguments for the biological, economic, and practical advantages to zoning the intensity of practice were advanced in British Columbia. Since then, practical experience has grown. Some companies have embraced the principles of hierarchical planning noted above, which help them to accrue economic benefits while still practicing a more 'gentle' approach to forestry.

Canada has long been pre-eminent in the field of habitat and forest modelling, and British Columbia is a veritable hotbed of such modelling, with a wealth of experience in developing and using models as well as evaluating them. Well used, such models are essential to the planning of forestry, selection of management regimes, and the improvement of forest practice, and are an integral part of an effective adaptive management program. It appears that the most completely developed and successful adaptive management programs for forestry have been implemented in British Columbia, despite—or possibly, because of—barriers imposed by rule-based legislation. Learning how to learn while manipulating a complex system with inherent time lags will become increasingly important as societal demands on forestry

---

8   That conservation forestry can be effective is evident in the receipt by Weyerhaeuser (BC) of the Ecological Society of America's Corporate Award 'In recognition of the Forest Project approach to Ecosystem Management'.

continue to change. We view effectiveness monitoring as part of adaptive management. A few companies in British Columbia have developed or are developing sound effectiveness monitoring programs that encompass the complexity of biological diversity. We are aware of no comparable examples elsewhere.

In summary, the economic returns from forestry within the province will continue to decline, but the province is well experienced in features of forestry that are likely to become increasingly important. It is unclear whether the struggle for market share will reward that experience, or whether there is sufficient creativity to otherwise turn that experience to economic advantage.

Lessons learned

Much of the scientific basis underlying the following issues are addressed in two recent books—*The living dance* (Bunnell & Johnson 1998) and *Forest fragmentation: Wildlife and management implications* (Rochelle et al 1999). We have chosen to address the subject in terms of lessons learned over the past 10 years, and believe there are seven large lessons.

1 Forestry is not rocket science—it is much more complex
Effective practice of forestry embraces the relatively tidy disciplines of physics, chemistry, and genetics; disciplines that consider large areas such as hydrology and meteorology; and the fuzzier fields of study such as economics, visual quality, and conservation biology. In short, it is probably the most complex endeavour humans pursue. Most practicing foresters know that; most of the public and legislators do not. Caught amid the swirl of continual change, it is unfortunate that foresters must accept one more task—that of informing legislators and the public of the nature of the foresters' charge. The alternative to the task is worse, and typically emerges as simplistic, rule-based approaches to forestry. Not only do these not work, but they severely constrain learning. Constraining the learning of those pursuing humanity's most complex endeavour is unhelpful.

2 Act on what you know, rather than what frightens you
British Columbia's rule-based approaches emerged from fears rather than science. It is true that most existing research findings

are not organised in a fashion that connects directly to the complexity of forestry. Performing such syntheses to expose relevant connections is time-consuming, but well worth the effort. Not only do such syntheses act as a form of due diligence, but they avoid the barriers imposed on future learning by simplistic rules. Moreover, the time required for such syntheses and focused research is minuscule compared to most rotation lengths. Computer simulations[9] indicate that it takes at least two rotations to alter a landscape pattern once it has been imposed on a forest. That observation is compelling enough reason to rely on science to guide practices to the extent that science can. For at least five years, the Forest Practices Code of British Columbia dictated landscape patterns that even the most simple reading of the principles of conservation biology would have revealed were undesirable. It will require at least 200 years to alter those mistakes. The mistakes were admittedly unintended and well intentioned, but ignored available tools and science. Conducting informative research within forests faces special challenges,[10] but employing scientific findings not only reduces practitioners' difficulties, it is the clear moral and professional choice. Some commitment to structured learning is essential. That includes future learning, as through adaptive management where monitoring outcomes can refine future practice.

3 Don't do the same thing everywhere

Given the complexity of forestry, and the enormously diverse array of species that dwell in forests, there should be no need to make this argument.[11] Unfortunately the need is real. There appears to be little appreciation that imposing stand-level practices with fine-scale variability repeatedly on a forest ultimately yields broad-scale homogeneity. Such homogeneity eventually compromises the sustenance of biological diversity and other future options desired from the forest. Our experience is that the most promising approach to sustaining all desired forest values, including economic returns, is through some form of structure-based management such as variable retention that sustains important structures on the

---

9   See pages 272–93 in Rochelle et al 1999.
10   See Bunnell et al 1999.
11   See reviews in Bunnell et al 1999 and Bunnell & Johnson 1998.

harvested land base. For these stand-level actions, 'variable' and 'retention' should receive equal emphasis. Stand-level practices must be implemented within a planning framework that creates the context for those actions. Strategic decisions (eg degree of representation across ecosystem types, potential restoration of old-growth attributes) and tactical decisions (eg sustained provision of dead wood, protection afforded riparian systems) provide context for the amount and form of retention during operational planning. Zoning the intensity of fibre extraction at the strategic level is very helpful, whatever silvicultural system or harvesting approach is used.

4 Embrace the freedom of humility.

Twenty-five or more years ago, foresters in British Columbia enjoyed public confidence and credibility, and were warmed by the 'myth of the omniscient forester'. Younger foresters may never experience those pleasures. Now the confidence—some would call it arrogance—has been beaten out of foresters by the public, by the challenges, and by a Forest Practices Code that prohibited them from developing solutions to the challenges. But we are not speaking of being beaten into humility, but of embracing the freedom of humility. When you are no longer expected to know everything, as professionals you are at least expected to learn. It is possible to learn by doing, provided that an effective monitoring system is in place. We call that process adaptive management. It is difficult to implement in any system and forests offer particular challenges; nonetheless, it is possible and the system developed by Weyerhaeuser (BC Coast) has provided sufficient guidance to managers to inform decisions about both planning and practice.[7] We have found that linking an adaptive management system to criteria and indicators of either a certification system or an internal plan for sustainable forest management helps to make the effort both more informative and cost effective.

Given that society is going to continue to move the goalposts for forestry, sustainable forest management will remain a direction not an endpoint. That condition ensures that learning must continue. Provided that foresters can explain their field sufficiently well to evade simplistic rules, we are convinced that adaptive management is the best way to ensure learning. Adaptive management incurs costs in monitoring, in data management, and in experimental design. Nor is it easy to design well. However, the costs of not learning are much higher. Moreover, humility gains little if it merely leaves its owners bashful and quiet. The greatest value of genuine humility, beyond restful nights, is the opportunities it creates for joint learning.

5 Hold hands before crossing the street

We have learned in British Columbia that some of the traffic is dangerous, but there are plenty of folk with whom to hold hands. Two groups are particularly important—critics and researchers. Embracing critics with genuine humility engages them in the process of learning to do better, thus enhancing learning and reducing conflict. Like adaptive management, embracing your critics also incurs costs, but these are outweighed by the advantages. Meaningfully engaging researchers with practitioners is a still older problem. Much of the problem it stems from the different reward systems and senses of risks applicable to each group, which can engender lack of respect. Mutual respect for the different reward systems and senses of risk will be insufficient without a liberal increase in honesty. Researchers must be more forthcoming when they do not know how to answer a management question, and less eager to disguise their own curiosity as something of immediate operational significance. Practitioners must be less fearful of denying research that they have reasoned will not address their problems. Combined, these tactics would serve to increase the effectiveness of applied research and to reduce time lags between societal expectations and science to address the expectations. Such an approach will not provide applied researchers with the esteem enjoyed by curiosity-driven researchers, but it will make them more effective. An obvious question in the search for approaches to sustaining biological diversity is 'How much is enough?' In some instances, this question can be reasonably well answered by thoughtful syntheses or meta-analyses of existing data;[12] in other instances, synthesis is insufficient and an experimental approach is necessary. Cooperation or 'holding hands' is particularly advantageous during operational experiments.

---

12 For example, see Bunnell et al 1999 and Rochelle et al 1999.

6 The only advantage of 'Ready, fire, aim' is speed

In British Columbia the Forest Practices Code of 1995 attempted to cover all the complexities of forestry and was created in a seeming whirlwind of activity. Although there were many improvements to practice engendered by the code, the haste—or lack of respect for science—that prohibited review of the literature also created serious problems (some noted above), which led to what was intended as revision in the form of a results based code. Having shot themselves in the foot, the code creators, reloaded and aimed higher. The new code was created with more speed, and even less attention to science. We learned and are relearning that development of thoughtful regulations governing the sustenance of biological diversity requires time to think, synthesise, and analyse. That is particularly true when landscape patterns created by practices require at least two rotations to modify. Given that some legislators and civil servants are willing to ignore the costs to future generations, it may fall to researchers and practitioners to resist the haste that prohibits thought. Our experience also indicates that little can be done to introduce thought when the moral imperative is not there. In those instances, companies bear an additional burden besides focused adaptive management and monitoring necessary to improve their own practice. They also need to monitor the effectiveness of regulatory frameworks and their associated costs. In British Columbia, the government has significantly reduced its responsibility for public lands. Elsewhere, government agencies may be able to help companies with either research or monitoring.

Restart John Muir's sawmill

Although Abraham Lincoln granted the land for the first national park in the United States, John Muir (founder of the Sierra Club) was its spiritual parent. Muir also owned and operated a sawmill in Yosemite Valley, and he saw no conflict in setting aside the first national park and wilderness area, with a logging operation nestled within it. We subsequently ignored that approach to zoning and embraced multiple use with a fervour that saw all forest values theoretically sustained in a small area. Multiple use never worked well when the values were few. It does not work at all when the values are many and some are in direct competition. For one thing, multiple use leads to chronic rather than episodic disturbance, with a significant level of disturbance over most of the land base all of the time. Our experience is that zoning the intensity of forest, even within a single forest tenure, greatly facilitates planning and operations, and does much better at meeting societal expectations. True, we will not argue for restarting John Muir's specific saw mill, though there is ample evidence that some national forests of the United States could use a mill if they are to sustain their current values. We do believe that some form of zoning the intensity of forest practice is beneficial, particularly when combined with thoughtfully applied stand and landscape practices.

Acknowledgments

We are grateful to the granting agencies, federal and provincial government ministries, non-governmental organisations, and forest companies that have supported our efforts to improve sustainable forest management in British Columbia. We appreciate the contributions of Glen Dunsworth that improved this manuscript.

7

Finnish forestry in transition— finding ways to sustainable forest management

Jari Niemelä

Background: drivers of change in Finnish forestry

Finland is a country of forests. As much as 70 per cent of the country is covered by boreal forest. Consequently, forests are of great economic, ecological, and socio-cultural importance. As well as providing raw material for industry, forests are used for recreation and nature tourism. Furthermore, the native Sami people base their livelihoods on forests (reindeer herding, fishing, hunting). Forests are also an important environment for Finns as a source of spiritual inspiration and enjoyment. However, due to the intensive forestry, many species and biotopes are threatened in Finnish forests. Forests harbour about 40 per cent of the threatened species in Finland, and about one-third of the threatened species in the country are declining due to forestry practices. Furthermore, forestry is in conflict with the traditional Sami use of natural resources, and to some degree with recreational uses and nature tourism.

The traditional way of resolving the conflict between intensive forestry and nature conservation has been the establishment protected areas. In Finland, most of the large, and therefore ecologically and recreationally most valuable protected areas are in the north, where the forests are rather unproductive and mainly owned by the state. Up to 20–30 per cent of forest land is protected in the northern parts of the country, while only about 1 per cent is protected in the more productive and mainly privately owned southern forests.

During the past decades it has become evident that protection of forest biodiversity cannot be solely based on protected areas. In particular, this is the case in southern Finland, where protection is expensive and there is simply not enough ecologically valuable forests left to be protected. Therefore, a societal demand emerged, especially from environmental groups and also scientists, that forestry practices in managed forests must become more 'environmentally friendly' and take biodiversity into consideration. These demands were also voiced from countries with which Finnish forest industry does business, in particular other European Union countries. This situation lead to rapid changes in forestry practices in Finland, as well as in Sweden and Norway, during the 1990s. The previously dominant clearcutting has been modified by adopting the 'variable retention' approach which means leaving more live and dead trees in the harvested sites. Furthermore, the size of the cut-blocks (coupes) has decreased to some 2 hectares in southern Finland.

From an ecological point of view, the important question is whether or not the modified harvesting methods sustain biodiversity in our forests. There is no unambiguous answer to that question yet, but there is research going on to answer it.

From the wood production point of view, it is claimed that Finnish forestry is sustainable because the yearly growth of trees is more than the volume harvested. But sustainable fibre production is not enough for many interest groups in the society who advocate a more holistic view of sustainability. Consequently, the integration of ecological, economic, and socio-cultural sustainability is the current goal of Finnish forestry. To understand why the change towards an ecologically and socio-culturally sustainable forestry is taking place it is useful to have a look at the history of forest-related conflicts.

In Finland—as in many other countries—forests and forestry have been a battlefield for a variety of interests for a long time, but conflicts between different stakeholders have intensified since the 1950s and especially since 1970s. These conflicts emerged more or less simultaneously in many countries. The development leading to recent forest conflicts and consequent changes in forestry practices has three interlinked components: (1) forestry operations have intensified, (2) recreational needs have increased, and (3) the environmental movement has developed and gained strength. Thus, during the past decades forests have become objects of an increasing number of interests and interest groups having different and clashing views about forest management.

Striving to economic growth is often seen as the principal cause of the intensification of forestry. Intensive and efficient forest management yielded better economic returns and lead to increased living standards, and thus increased demand for recreational use of forests. On the other hand, the same forestry that produced more economic growth, and thereby wealth, was perceived to cause deterioration of the forest environment. Consequently, the environmental movement gained strength which was at least partly fuelled by scientific reports on environmental problems in forests. Thus, forest conflicts are intimately linked with social development in industrialised counties during the past decades with such phenomena as urbanisation, continued industrialisation and increased standard of living leading to changed values among people. This trend appears to be true in many countries worldwide.

Interestingly, some Finnish researchers claim that forest-related conflicts were partly caused—or at least aggravated—by the inability of forestry professionals to adapt their views and approaches to accommodate changes in the society. The previous criticism towards forest management had focused on biological issues (such as the biological effects of herbicide and pesticide use, and acid rain), economic, and technological issues. Therefore, forestry professionals did not have the tools to cope with the new type of criticism that, in addition to the old type of criticism, dealt with values, social issues, recreational demands, and the new concept of biodiversity. The new situation lead to confusion among forestry professionals, and consequently to an inability to find common strategy to deal with the new kind of criticism.[1] This state

---

1 See Hellström & Reunala 1995.

of affairs has changed, and significant changes have taken place in the Finnish forest management procedures during the past decade.

It appears that the situation as regards sustainable use of forest resources is in a very dynamic state in many countries. This is particularly true in Finland where a multi-stakeholder committee with the task to find ways of protecting forests in southern Finland just finished its work without reaching consensus among the parties. Environmental groups feel that too little was achieved in terms of forest protection, while forest owners and forest industry are relatively satisfied. The report of the committee focuses on the conflict between protection of forest biodiversity and forest harvesting. Much less emphasis is placed on the socio-cultural aspects of forest use, such as recreation and tourism. Thus, the main clash between the different stakeholders in the forest management discourse appears to be between forestry and protection of biodiversity.

Major policy and technical issues of Finnish forestry

The current state of forest discourse

It is difficult to pinpoint the causality of the main forest management issues and their drivers in Finland because they are tightly interlinked. We may have to wait another decade to find out what really happens today and why. The policy issues stem from the above described interlinks between societal changes and forest management. In a broad sense, important drivers of forest policy are public concern about the ecological consequences of forestry, and national and international agreements.

Finnish forest industry is highly international. Thus, policy issues come not only from within Finland but also from other parts of the world, especially other European Union countries. One of the pressures contributing to changes in the Finnish forest sector is the so-called 'market pressure', that is, consumers demanding changes in forest management by selecting the most environmentally friendly products. There are contradictory views about whether 'market pressure' is exerted by ordinary consumers demanding wood products from forests treated with environmentally friendly logging methods or is 'market pressure' created by vocal and strong environmental groups acting as 'representatives' of consumers. Nevertheless, environmental groups are one of the drivers of changes in Finnish forest management, for instance, by influencing buyer companies of wood-based products and thereby 'forcing' the producers to change forestry methods.

Perhaps the best-known European example of the significance of 'market pressure' and environmental groups is the declaration of the large German publishing house Axel Springer Verlag which sets strict environmental standards for their paper suppliers. These six Forestry Standards for the suppliers have been in force since 1995:[2]

1   Sustainability: Harvesting more timber than will re-grow is prohibited.

2   Biodiversity: Forestry shall not endanger animal or plant species.

3   Control: The paper manufacturer as a purchaser of timber must perform eco-controls.

4   Training. The paper manufacturer must ensure that the necessary ecological knowledge is made available to personnel (eg forest workers).

5   Indigenous population: The paper manufacturer must take the indigenous people (eg the Sami in Northern Scandinavia) into consideration.

6   Information: The paper manufacturer must keep the public informed of the advances made in environmental protection—but also of the problems encountered.

Another, and related, policy issue and pressure to change forest management comes from international agreements and consequent forest certification. Forestry and biodiversity were among the priorities of the UN Conference on Environment and Development in Rio de Janeiro in 1992 resulting in the Convention on Biological Diversity, the Forest Principles and a forest component of Agenda 21. In Europe, the implementation of these commitments has resulted, for instance, in the Ministerial Conferences for Protection of Forests in Europe (MCPFE) set up by the Forest Ministers of the European Union member states. MCPFE has developed a set of *Pan-European Criteria, Indicators and Operational Level Guidelines for Sustainable Forest Management* which also are implemented in

---

2   See http://www.asv.de/englisch/umwelt/frame.htm

current certification schemes. The Pan-European Forest Certification (PEFC) scheme is a voluntary, private-sector initiative that aims to provide assurance to the customers of forest owners that the products they buy come from independently certified forests managed according to the Pan-European Criteria as defined by the resolutions of the Helsinki (1993) and Lisbon (1998) Ministerial Conferences on the Protection of Forests in Europe. The Finnish forest sector has adopted a certification system of its own (FFCS, Finnish Forest Certification System) which is in accordance with the PEFC certification.

The third major policy issue of Finnish forest management is legislation. In both Finland and Sweden, forest legislation has been revised during the 1990s. From the point of view of ecological sustainability, the Swedish legislation is stricter stating that biodiversity and economic considerations have equal weight in planning of forestry operations. In other words, planners must seriously take biodiversity into consideration. The Finnish law does not emphasise biodiversity as strongly, but requires biodiversity to be considered in forestry operations. For instance, the legislation includes a list of 10 forest biotope types (so called 'key habitats') that must be set aside whenever encountered. These are usually small, but ecologically significant and rare types of biotope patches which are expected to host red-listed species. Although such sites must be left untouched, there is no guarantee that these sometimes isolated biotope patches maintain their characteristics and species.

Fourthly, in addition to signing international agreements, Finland has produced a national forest strategy which aims at ensuring sustainable forestry in the country. The strategy comprehensively treats the different components of sustainability (economic, ecological, socio-cultural) and sets targets for the future. Although the document was produced by intensive and long-lasting consultations and participation by various stakeholders, some feel that the strategy is too economy-oriented and emphasises timber production at the cost of other components of sustainability.

How have these issues been tackled and resolved? Landscape ecological planning as a tool
All the above mentioned drivers have contributed to producing clear changes in how forestry is conducted in Finland. For instance,

certification requires that biodiversity is taken into account in forest management and there are requirements that must be met for receiving the certificate. Legislation has been revised to accommodate the new demands of sustainability. In order to help forest owners comply with the new legislation and agreements, various forestry organisations have prepared guidelines. The private forest owners' organisations are very efficient at covering basically all forest owners and providing guidance in matters related to forest planning, logging methods, and biodiversity considerations.

Although 75 per cent of forest land is privately owned in Finland, the state is the single largest forest owner and therefore a significant timber seller. The forest land owned by the state is managed by the Finnish Forest and Park Service. During the 1990s the Forest and Park Service has revised its forest planning system in order to reach the goal of sustainable forestry. The system is called 'landscape ecological planning' and it now covers all the silviculturally managed forest land (about 6 million hectares) of the Forest and Park Service. There are more than 100 planning areas, each ranging from a few thousand hectares to over 100,000 hectares. The aim is to make sure that forests are managed sustainably from an economic, ecological, and socio-cultural point of view.

In terms of ecology, the goal of landscape ecological planning is to ensure that logging in the planning areas does not result in species losses, that is, the natural biodiversity of the area should be maintained. Scientifically, the method is loosely based on the island biogeography theory including emphasis on ecological corridors.

The planning system includes two levels: (a) the landscape level, and (b) stand level. The planning method requires knowledge of the ecologically valuable sites in the planning area, such as old-growth forests, springs, and brooks. Ecologically valuable sites are the ones included in the forest law, but in addition, the Forest and Park Service also defines some other site types as valuable. In the planning, the valuable sites are set aside and ecological corridors are created to connect larger patches of high-value forest or existing protected areas. Prescribed burning is used to create habitat for pyrophilous species, and restoration is practised to re-create habitats.

On the stand level, variable retention is used. This means that groups of trees and single trees are left in the cut-blocks. The retained trees include old and dead trees, and groups of trees in moist patches or sites difficult (and therefore expensive) to harvest. This kind of variable retention harvesting system results in quite a different appearance of the cut-blocks as compared to the traditional clearcutting. For instance, a considerable amount of coarse woody debris is left.

Participatory planning involving local stakeholders is used in the Forest and Park Service's landscape ecological planning. Various stakeholders, such as environmental groups and local residents, are involved in the planning process. This is a new approach, and requires further development.

A recent assessment by external evaluators concluded that the Forest and Park Service's landscape ecological planning system is a significant step towards sustainable forestry, but whether or not the goal of not losing biodiversity values in the management areas is reached could not be assessed in detail. However, the probability of biodiversity being maintained is increased as about 10 per cent of forests in the planning areas will be entirely or partially set aside. In addition to areas set aside and thereby left entirely or partially unharvested, the amount of retention trees (both dead and alive) will be increased in harvested stands. For instance, the amount and diversity of dead wood will increase as a result of the planning system being in place. Dead wood is an important substrate for many species, but the amount of it is very low in Finnish managed forests due to very intensive forestry. The goal of the Forest and Park Service's landscape ecological planning is 5–10 m³/ha in ordinary forests and 1020 m³/ha in ecologically valuable sites. The lower goal has been already reached.

In terms of economy, the decreased logging possibilities caused a 10–12 per cent reduction in economic output during the first 10–20 years as compared to a harvesting alternative in which only the restrictions posed by the forestry law are considered. However, later on the decrease in logging possibilities is much less.

Regarding sustainability of forestry as a whole, the evaluators of the landscape ecological planning system pointed out that the economic and ecological considerations have received most attention in landscape planning, while socio-cultural sustainability has played a minor role. Consideration of the socio-cultural sustainability is more difficult but it is becoming increasingly important.

In conclusions, the experiences from the Finnish Forest and Park Service's landscape ecological planning are encouraging. Currently, the planning system is being improved based on the recommendations by the evaluation group. In private forests, however, landscape level planning is more difficult. There are about 400,000 forest owners in Finland, and the average size of a forest holding is some 30 hectares. Thus, it is a great challenge to take the diversity of ownership patterns into consideration when planning at the landscape level in private forests. On the other hand, new forest harvesting guidelines and certification ensure that in private forests biodiversity is also taken better into consideration than before.

In Finland, a hotly debated national policy issue is the protection of forests in the southern part of the country where most forests are privately owned. In the south, only about 1 per cent of the forest area is protected, which is clearly not enough for the maintenance of forest biodiversity. Environmental groups, in particular, have demanded more protected areas to be established in the southern part of the country. The way that this problem was tackled was to establish a multi-stakeholder committee as mentioned above. However, the issue remains largely unresolved, as the committee was not able to reach consensus about how much forest should be protected and how the society should go about protecting it.

Research is needed to support the changes
In order to successfully resolve problems of sustainable forest management, scientific understanding of the issues is needed to provide solid understanding to support decision-making. In Finland, research has been important in providing a framework for changes in forestry. Numerous studies on various aspects of forest biodiversity and the effects of forestry practices on it have been conducted. The multidisciplinary Finnish Biodiversity Research Programme (FIBRE)[3] has been instrumental in funding and coordinating such research. This six-year program (1997–2002) has funded about 15 research projects dealing with forest biodiversity.

A special project with the aim to synthesise research results for the needs of practical forestry was included in FIBRE. This project

---

3 See http://fibre.utu.fi/

has organised meetings, workshops, and field visits to facilitate communication and collaboration between researchers and (other) stakeholders. The project has been successful in its task, and it seems that lasting partnerships between researchers and end-users have been established.

In addition to national research, the European Union is funding research projects dealing with sustainability of forestry. The EU-funded project 'Indicators for monitoring and evaluation of forest biodiversity in Europe BEAR' that ended in 2001 was one such initiative,[4] and currently the BIOFORUM project[5] deals with biodiversity conflicts, among others in forests.

Several scientific congresses have been organised in Fennoscandia during the past years to discuss issues of forest biodiversity. The proceedings of these congresses form a valuable set of scientific contributions that also have useful information for forest managers.[6]

An example of the kind of research that is useful for forest management is that done on the 'ecology' of coarse woody debris. It has been shown that numerous species of various taxonomic groups depend on coarse woody debris as habitat. In Finland, about 20 per cent of the 20,000 forest-dwelling species live in dead wood. In addition to revealing numbers of species dependent on dead wood, research has established that there is a positive, but non-linear relationship between species richness and the amount of coarse woody debris at the forest stand level. This relationship can be used to set quantitative goals for retention of dead wood.

Natural disturbance dynamics of forests is another example of a research issue with practical applications. For instance, the Finnish Forest and Park Service uses the natural disturbance regime of forests as a guideline in their forest management. The idea is that forests that burn frequently in natural conditions (eg dry pine-dominated forests) could be harvested more intensively as their biota are used to frequent and severe disturbances. Forests that burn infrequently (eg moist, spruce-dominated forests) would be left untouched or harvested lightly.

---

4 See Larsson 2001.
5 See http://www.nbu.ac.uk/bioforum/
6 See Larsson & Danell 2001.

Has forestry changed in Finland?
Although Finnish forest management has evidently become more 'environmentally friendly' in the past decade, the question of whether or not forestry has changed enough to ensure the maintenance of forest biodiversity remains partially unanswered. Some feel that changes are not sufficient to guarantee species survival in Finnish forests, and call for more protection. However, all the stakeholders agree that the new management procedures and approaches are a step towards ecologically sustainable forestry. For instance, the landscape ecological planning system of the Forest and Park Service increases the probability of forest-dwelling species surviving in the planning area as compared to earlier, clearcut based forestry.

There is a clear change in the views and attitudes of forest professional in Finland. Today, there appears to be true interest among forest planners, managers and forest owners to protect biodiversity. This is reflected in the willingness by at least the large forest owners to voluntarily set aside valuable sites. An example is the protection of key habitats. According to an inspection conducted in 1999, 25 per cent of all the ecologically valuable sites set aside by the Forest and Park Service were key habitats defined in the forest law, while 75 per cent of them were habitat patches defined as valuable by the organisation's own guidelines. Even private companies are setting aside more forest than the law requires. In addition to setting aside 8000 ecologically valuable sites as defined by the law, the forest company UPM-Kymmene will voluntarily protect another 12,000 valuable sites. These two examples show that large forest owners actually set aside considerably more sites voluntarily than is required by the law.

What might have been done better?
It is difficult to define any single matters that could have been done better in the Finnish forest protection discourse. However, one area that could have been better handled is that of forestry-related conflicts. These may be impossible to resolve completely, and a certain degree of conflict may even be beneficial for keeping all the stakeholders alert. The problem is that in Finland, forest-related conflicts have seriously polarised the discussion about sustainable forestry. These conflicts mainly occur between environmental

groups who are worried about biodiversity being lost due to forestry and forest industry and owners who see their businesses or livelihoods threatened by more forest protection. There are also conflicts between nature tourism and forest industry. This is because nature tourism depends on large, pristine forest areas that tourists and recreationists want to visit and enjoy.

In retrospect, it appears that the conflicts have polarised the discussion too much, which has delayed progress in reaching solutions on the sustainable use of forest resources. It is understandable that environmental groups become frustrated as they see forests that they consider valuable being harvested. On the other hand, forest managers feel that their work to sustain timber production and the economic wealth of the nation, and, more recently, their efforts to maintain biodiversity are not appreciated.

Future directions

The future is difficult to predict, in particular the future of a dynamic discourse in the society, such as the forest discourse in Finland. Issues may change rapidly and their causal relationships are difficult to disentangle. However, there appear to be some general future directions that are likely to persist. One of them is the drive towards truly multipurpose forestry aiming at overall sustainability. So far, the discussion in Finland has focused on the ecological effects of forestry and ways of mitigating adverse effects on biodiversity. This is also reflected in research. An impressive amount of research on ecological issues and forest biodiversity has been published recently.[7] Undoubtedly, ecological research has advanced with giant leaps during the past decade and there is now a fairly solid ecological basis for decision-making in forestry.

The other components of sustainable forestry (economic and socio-cultural) have not received as much attention as ecology/biodiversity in the forest debate or in research. However, this state of affairs is changing as the debate on issues such as the relationship between logging and recreation/nature tourism or between logging and traditional reindeer herding is gaining momentum. Also research on these issues is increasing. For

---

7 See Jonsson & Kruys 2001.

instance, a sizeable research project entitled 'Forests and good life' funded by a private forest research foundation has just been launched. The aim of this research is to examine the links and trade-offs between the various components of forest sustainability.

Another future direction relates to the scale of forestry planning. It is evident, partly thanks to research, that planning must be done on the landscape level in order to maintain forest biodiversity. As mentioned above, the Finnish Forest and Park Service has established a landscape ecological planning system which is currently being developed further based on the recommendations of the evaluation. The tools developed for planning in state-owned forests covering large forest areas are now being adapted for use on company lands and small forest owners' lands.

Related to the planning of forest management is the development of monitoring methods and indicators for changes in forest biodiversity. This issue is of considerable international interest as well. Recently, the BEAR project completed its work on the issue and published a comprehensive treatment on forest biodiversity indicators in Europe.[8] In Finland, a national process involving the main stakeholders has been launched to develop monitoring and indicators for forest biodiversity.

Overall, research is very important in providing scientific understanding and a basis for decision-making. During the past decade, a significant amount of ecological research has been conducted in Finland to document biodiversity losses due to forestry and to find ways of mitigating the adverse effects. This kind of research will continue but also multidisciplinary research focusing on the overall sustainability of forestry will increase.

An example of the kind of practical research needed to support decision-making in forest management is the studies on the usefulness of the so called 'key habitats'. The Finnish legislation defines 10 types of key habitats in forests that must be left untouched when logging. These are small forest stands or habitat patches that can be expected to host red-listed species. It has been estimated that these habitats cover only 0.5–3 per cent of the productive forest land, but because of their characteristics, they may be very important for forest biodiversity. Although the main-

---

8 See Korpilahti & Kuuluvainen 2002.

tenance of key habitats in the managed forest landscape is potentially good for biodiversity, several issues must be considered and examined before the concept is fully operational. For instance, it would be best if key habitats could serve as functional ecological units that maintain entire communities. Furthermore, the spatio-temporal dynamics (including successional changes) of stands and habitat patches need to be considered in forestry. Leaving single forest patches here and there may not efficiently serve biodiversity conservation.

At the societal level, there are processes in Finland attempting to ensure the sustainability of forestry. As mentioned above, the multi-stakeholder committee to develop ways of increasing the far too low proportion of protected forest area in southern Finland suggested several ways of increasing the level of protection. The methods proposed by the committee are mainly voluntary actions based on the assumption that ownership of the forest does not change. For instance, the state may 'rent' valuable forests from private land owners to be protected for a certain time period. The land owner is compensated for lost logging possibilities and does not lose the ownership of the land. These approaches appear promising, as one of the main criticisms from private forest owners is that protection means losing the ownership of their forests which, in many cases, has been in the same family for centuries.

Important lessons

There are several lessons to be learned from the Finnish experience. First, rather rapid and relatively profound changes in forestry practices are possible. In Finland, major changes towards ecologically sustainable forestry took place during the 1990s. For instance, in just six years the Finnish Forest and Park Service prepared more than 100 landscape ecological plans covering 6 million hectares. However, it is too early to conclude whether or not the plans guarantee the maintenance of biodiversity in the managed Finnish forests.

Second, in order to produce a change, citizens must actively demand them. As 'environmentally friendly' forestry measures are costly, there must be a 'market pressure' to prompt forest industry to change practices. In Europe, such pressure is evident and propelled by vocal environmental groups. There are examples in Finland of the significance of the pressure. Forest owners who have logged in forests considered by environmental groups as ecologically valuable have not been able to sell their timber because industry did not want to buy the harvested timber for fear of bad publicity.

Third, it is vital to support research on sustainable forestry because research can provide understanding to support decision-making. Furthermore, it is important to create an environment and mechanisms for a dialogue between the scientific community and forest industry. In fact, communication among all the stakeholders in the forestry discourse is of vital importance.

Research and monitoring are very important, as in many countries the forest industry has already adopted new forest management methods that are likely to preserve biodiversity better than the old methods. However, no one can say whether or not the methods and approaches are working. That is, it has not yet been demonstrated that the rate of species disappearance has stopped or at least slowed since these policies were instituted. It is important that the modified harvesting methods be tested and monitored, for the sake of both industry and conservation. The expense to industry of instituting these policies is large, and industry deserves to know if they work. From the standpoint of conservation, the stakes are equally great. Society wants biodiversity maintained, and the public has been told that industry is doing many things that will help to achieve that goal. The only way to determine the efficacy of the new forestry is through controlled, replicated, large-scale experiments combined with monitoring. Multidisciplinary teams of scientists are needed to do this research.

Fourth, adaptive management is an approach that could be applied in the multipurpose use of forests to foster communication and collaboration between forest managers, other stakeholders, and scientists. Adaptive management can be described as 'learning by doing', and constitutes an evolution from conventional wisdom and best current data approaches towards an adaptive process including monitoring of biodiversity and modification of management practices. Adaptive management is a particularly useful approach when actors are forced to make decisions based on incomplete knowledge and when decisions are embedded in a world of uncertainty. This is usually the case in forestry, where there is uncertainty about the sustainability of the planned forestry operations, in particularly in ecological and socio-cultural sense. In a sense, Finnish forest managers are already using such an approach, but there is room for improvement. For instance, the Forest and Park Service uses participatory planning and the plans are revised every five years.

Fifth, in addition to the forest industry, it is important to consider the view of private, small forest owners. They may be economically dependent on the timber from their forests, and may thus have fewer options to practise 'ecological' forestry than do big forest companies. Maintenance of biodiversity in private forests is becoming an increasingly important issue in Finland.

8

The role of science in the changing forestry scene in the USA

David Perry

Scientists in the United States have become increasingly involved at many levels in the movement toward sustaining biodiversity within managed forest landscapes. Much of this effort (but not all) has been directed to federal lands. From an ecological standpoint, the challenge is precisely that facing basic ecology: to better understand the relationship between structure and function at a variety of spatial and temporal scales. In forestry that strives to maintain diversity and functional integrity, economic viability and social approval are added to the mix. The experience in the US shows the old model of scattered reserves within a sea of intensively managed forests doesn't satisfy these diverse goals, and foresters in the US, as in other places in the world, are re-examining basic assumptions and approaches. The newly formed National Commission on the Science of Sustainable Forestry (NCSSF, or 'the Commission') marks an evolutionary step in that for the first time private foundations have committed substantial funds to foster the role of science in developing diversity-friendly forestry on both public and private lands.

A brief overview of forests and forestry in the US

At present, 33 per cent of the US is forested (at least 10 per cent tree cover), a little over two-thirds of the estimated cover at the time of European colonisation. Forest area is divided approximately evenly between the predominantly coniferous forests of the west (including Alaska) and hardwood, conifer, and mixed forests of the east. More than 25 per cent of forest land, mostly in the arid or montane west, is not considered potential 'timber' land because of low productivity.[1]

Ownership patterns differ widely between the west, where nearly 70 per cent of forests are in public ownership (mostly federal), and the east, where nearly 90 per cent are privately owned. By far the majority of private ownerships are non-industrial (83 per cent in the east and 67 per cent in the west), including many small forested parcels that are part of farms or rural, non-farm land holdings. Sixty per cent of the nation's domestic roundwood came from non-industrial private lands in 1996, and 28 per cent from industrial lands. In terms of growing timber, the south and the mesic temperate forests of the Pacific northwest are the prime locations. Since heightened species protection on federal lands in the west (beginning in the early 1990s), the private lands of the south have increasingly become the fibre baskets for the nation. As of 1996, nearly two-thirds of timber harvest (volume) came from the south, with the rest divided evenly between the west and the north (upper midwest and northeast).

'Strictly' protected areas (mainly parks and designated wilderness areas) are concentrated on public lands, and therefore found predominantly in the west. Roughly 10 per cent of public forest land in the west and 3 per cent in the east fall in this category. The heaviest protection occurs in montane and boreal forests with the least commercial value, although since the early 1990s protection of public forests within the highly productive Douglas-fir region has increased significantly. The National Council for Air and Stream Improvement (NCASI), an industry sponsored research group, reports that less than 28 million hectares of private forest land in the US is managed under some form of voluntary program

---

1 Unless noted otherwise, statistical information in the first three paragraphs of this section is from The Heinz Center 1999 or from Johnson 1997.

'that promote(s) sustainability and stewardship'. Because such programs are diverse in objectives and requirements, and some are still evolving, it is difficult to generalise about what that implies for conservation and sustainability in the broad sense.

Changes in total forest area since colonisation do not reflect major alterations in the structure of the nation's forests. With the possible exception of some types, most forests at the time of colonisation were likely to have been 'old-growth', a term that means different things in different forest types (and doesn't exclude the presence of young trees), but nevertheless implies an age distribution and structural complexity that differs from younger forests, especially those managed primarily for fibre. At present, 55 per cent of potentially commercial forests are less than 50 years old, and only 6 per cent are older than 175 years.[2] Today's younger forests, many of which have established on abandoned farms (especially in the east) or following clearcutting, don't have the rich structural legacies left behind by natural disturbances. Fragmentation is a problem, especially in the east, where many stands exist as small islands within an increasingly urbanised landscape (with consequent heavy impacts on, among other things, neotropical migrant birds). An extensive road network alters hydrology (especially in mountains when coupled with clearcutting), adversely affects stream ecosystems, and provides ready access routes for pathogens, weeds, and predators (including humans).

Virtually all forests in the US are composed of native species; however, there have been widespread shifts in composition, some triggered inadvertently (eg due to fire exclusion), some purposely (eg replacement of longleaf pines by faster growing loblolly and slash pine in the south). Noss and Peters list nine forest types among the 21 'Most-Endangered Ecosystems' in the US.[3]

The shifting role of science in forestry in the US

Between the late 1940s and the late 1960s, science in US forestry was almost exclusively focused on two things: honing the edge of the industrial plantation model (ie maximising economic produc-

---

2 USDA 2000.
3 Noss & Peters 1995.

tivity) and on US Forest Service (USFS) lands and watersheds (because of the central charge to the agency to steward the nation's water). Beginning in the late 1960s and early 1970s, various factors came together to dramatically broaden that role.[4] Research fostered by the International Biological Progam (IBP) (continued under the Long Term Ecological Research program (LTER)) revealed old-growth forests as vital and ecologically rich rather than the biological wastelands they had often been portrayed. Studies of the recovery of Mt St Helens, and of factors underlying degradation of high elevation clearcuts, demonstrated the crucial importance of biological legacies in ecosystem recovery (a new concept for ecology as well as forestry)[5] raising consciousness about the importance of what was left as well as what was taken during harvest. Second rotation decline in Australia and New Zealand raised red flags in the US about treatment of soils and led to inten-sified research on long-term site productivity. The US Congress during those years was in a mode of protecting diversity and the environment, passing among other significant legislation the National Forest Management Act, which directed the USFS to seek scientific advice about maintaining native diversity on USFS lands. A number of analyses and reports by USFS biologists followed closely on the heels of the act, some of which laid out principles and approaches for diversity-friendly forestry.[6]

The emergence of the spotted owl as symbol of old-growth-dependent species, coupled with the courts' insistence that federal land management agencies obey the law, catalysed the next phase of scientific involvement, which went beyond analyses to derive management alternatives for federal lands within a given region and rate each with regard to probability of meeting diversity goals. Several of these have been carried out in the western US (and one in western Canada that I'm aware of), but FEMAT (Forest Ecosystem Management Assessment Team, dealing with federal forests in the spotted owl region) stands in my mind as the exemplary model (albeit costly in terms of scientific time and effort).

---

4   Briefly reviewed in Perry 1998.
5   Franklin 2000.
6   The most thorough and best known of these was the so called Blue Mountain Book, Thomas et al 1979.

Over the past 10 years, scientists in the US and Canada have continued and deepened their involvement in analysis and advice regarding forest policy, at scales ranging from single landowners to entire regions, to all USFS lands in the nation.

The National Commission on the Science of Sustainable Forestry

The NCSSF was formed in 2000 under the auspices of the National Council for Science and the Environment, a 'policy neutral, nonprofit organisation dedicated to improving the scientific basis of environmental decision-making'. The Commission's formation was catalysed by four charitable foundations[7] who pledged support for a program aimed at:

- identifying key scientific questions related to sustainable forestry
- stimulating and overseeing research addressing those questions
- fostering a program of outreach and education to '... ensure research is transformed into meaningful results'

'Sustainable' is used here in the broad sense of sustaining all values that flow from forests, including biodiversity, water, productivity, and social capital; however, the initial charge was to focus on biodiversity (Montreal Process Criteria 1). NCSSF is an evolutionary step for at least two reasons. First, although private foundations in the US have been very active in conservation and environmental issues, NCSSF marks the first time I'm aware of that they have committed significant resources to the broader goal of science applied to sustainable resource management. Second, following logically from the first, NCSSF has the broadest charge to date, essentially all forest lands of all ownerships in the nation.

The Commission was kept lean and designed to deliberate efficiently and act quickly, a considerable challenge given the large academic component and the mixing of scientists and managers,

---

7   The David and Lucile Packard Foundation, Surdna Foundation, National Forest Foundation, and Doris Duke Charitable Foundation

but pulled off nicely thanks to focused leadership and a generally cooperative spirit among commission members. Based on recommendations from academia and various stakeholders (environmental organisations, industry), 17 members were chosen, divided evenly between a scientific panel and a stakeholders panel, with a chairman as the 17th member. The Commission was rounded out by a program director and small staff. Members of the scientific panel were drawn from academia (5), a conservation NGO (1), and industry (1), while the stakeholder panel represented environmental NGOs, public land managers, industry, and nonprofit policy analysis.[8]

In a series of three 2-day working sessions, supplemented by networking, the Commission developed first-stage research priorities that boiled down to:

- synthesising, either from published research or documented field experience, what is known about various aspects relevant to the Commission's mission

- leveraging ongoing research programs in order to broaden their scope and hasten the delivery of relevant results to managers and policy makers

  stimulating the development of support tools

Over a period of several months, proposals were solicited, screened by commission members, and reviewed by outside reviewers. Eight were funded in the first round. Without going into detail, these were:

- a synthesis of science connecting forest diversity/complexity to hydrology and aquatic ecosystems

- a survey of lessons learned from practical experience

- an assessment of knowledge concerning how management for non-timber products influences biodiversity

---

8 Membership as of July 2002: Chris Bernabo, Director; Norm Christensen (Duke Univ.), Chair; Ann Bartuska (Nature Conservancy); Jim Brown (Oregon State Forester); Chip Collins (The Timberlands Group); John Gordon (YALE Univ.), former Chair; Al Lucier (NCASI); Ron Pulliam (U. Georgia); Al Sample (Pinchot Inst.); Mark Shaffer (Defenders of Wildlife); Joyce Berry (U. Colorado); Bruce Cabarle (WWF); Wally Covington (N. Arizona U.); Sharon Haines (Int. Paper Co.); Dave Perry (Oregon State U. & U. Hawaii); Hal Salwasser (Oregon State U.); Mark Schaefer (NatureServe); Tom Thompson (USFS); Phil Janik (USFS), member during 2000 and 2001.

- an assessment of research needs associated with fragmentation effects

- an assessment and recommendations for core biodiversity indicators

- an assessment of needs and requirements for decision support systems

Two projects, each representing quite different regions of the country (the Southeastern and Pacific Northwest) that relate stand structures or landscape patterns (ie the spatially explicit mix of stand structures) to the diversity of key indicators (species potentially sensitive to management); and evaluate productivity trade-offs associated with protecting diversity. Of all projects funded by the Commission, these were the most directly tied to place, and are discussed in more detail in the section on place-based research below.

Obviously there was a heavy emphasis in the first round on assessing and synthesising what is known. There is no doubt about the wisdom of taking stock at the beginning, but this emphasis also reflects time constraints imposed by the funders and differing views within the Commission. Predictably, the funders wanted quick results. Some commission members felt most of the knowledge needed to protect biodiversity was already available and the highest priority was to pull it together and make it accessible to managers. I disagreed with that view (and still do), recalling one of the major conclusions of an NRC[9] panel convened in the early 1990s to assess the state of forestry research in the US: '...'the existing level of knowledge about forests is inadequate to develop sound forest management policies.'[10]

Nevertheless, the Commission ended up probably right where we should be in the initial phase, which is putting the competing beliefs of inadequate versus adequate knowledge to the test, or at least the partial test of literature and experience to date. (As with all things having to do with sustainability, the true test of whether we have it right or not comes only with time.)

---

9 The National Research Council, an arm of the National Academy of Sciences, routinely pulls groups of scientists together, at the behest of the US Congress, to assess and report on various issues.

10 National Research Council 1990.

The importance of place

Although it may seem like painful elaboration of the obvious, it's worth restating why general questions about protecting biodiversity must be answered in relation to particular places (e.g. forest types or regions). At least three reasons come to mind. First, scientists and managers need a better understanding of the limits of generality, or looked at in the inverse, how far one can run with a given model. For example, what does information concerning the use of corridors in one region tell about other regions? Second, realistic biological and social options may vary widely, as between the US west with some remaining old-growth and lots of public land, and the US east with little public land or old-growth. Third, it seems unlikely to me that managers will take seriously schemes they do not see as reasonably grounded in their particular place (wisely so, I expect). The latter two points in particular flow from the fact we are not just trying to learn something, we are trying to learn things that will be used.

So, what do we need to know about a given region in order to sustain (and perhaps restore) its diversity? A complete list would include:

1   the special places that warrant reserve status

2   the species or guilds of species likely to lose habitat under an intensive forestry (fibre-focused) regime

3   approaches to stand management that preserve endangered habitats (diversity-friendly)

4   the probabilities of sustaining regional diversity under various combinations (amounts, patterns) of landuses (reserves, fibre-friendly, diversity-friendly)

5   the implications of different landuse scenarios for carbon sequestration (where economics is a central consideration, as in the southeast US, managers are unlikely to undertake diversity-friendly approaches without knowing the cost in terms of productivity, and carbon markets may emerge as a way to help protect diversity)

6   the risks associated with different landuse scenarios, especially with regard to fire, wind, insects, and pathogens

Others may wish to add to this list, or to argue that we really don't need to know all these things. I think we do need to know all, or at the least be aware of the need and make some good guesses even if hard data are not available.

The most straightforward of this list is probably the first, identifying special places, as extensive databases exist from organisations such as the Nature Conservancy (at least in the US). Such data probably don't touch all the bases needed for an effective conservation reserve system, but they are an important start.[11] In the first round, NCSSF focused on numbers 2 to 5, with number 6 to be included in a future round.

There was a fundamental problem at the outset. While there are many examples of industrial forestry, there are few of silvicultural approaches that maintain greater structural diversity, and until the last 15 years virtually no experimental comparisons. Therefore we had to take a page from the astronomers and geologists, encouraging researchers to use the available raw material to look retrospectively at what happened. Here is what we asked for in the proposal solicitation:

Project 1. Diversity at the stand level.

For major forest regions, document the relationship between structural diversity at the stand level and populations of key indicator species, with the objective of testing hypotheses about the relationship between diversity (at the stand level) and silviculture systems spanning a range of forest structural diversity (this may entail developing an index of structural diversity). Include natural forests as reference areas, and within natural forests, elements unlikely to be found in existing managed forests (eg diversity associated with large decaying logs). Include soil biodiversity and treatments such as intensive site preparation. Assess the quality of existing information, synthesise what is known, identify potential key indicator species, identify gaps in knowledge, and conduct field measurements to fill those gaps. To the extent possible, field work should utilize existing

11 The Sustainable Forestry Initiative, an industry sponsored certification program in the US and Canada, has recently committed to use NatureServe data bases (from State Heritage programs) to identify—and presumably protect—ecologically special places on member industry lands.

variability in managed forest structure (eg spacing trials, vegetation management experiments, selection silviculture, ecosystem-centered management). Where existing treatments are not sufficient to cover the range of possibilities and give scientifically credible information, suggest experimental manipulations to fill the gaps.

Project 2. Assess tradeoffs between incorporating diversity and crop productivity.

This is the first step in an economic assessment of managing for diversity (later steps should include factors such as the influence of diversity on the spread of disturbances). Assessments must be conducted (a) across stands that exhibit a range of structural diversity (stands used in Project 1 might suffice), and (b) in which crop trees are near rotation age. (The latter requirement is critically important because experience with longer-term experiments shows that early response patterns may not be good predictors of what happens later on).

Project 3. Diversity at the landscape/regional level.

Select areas of high conservation priority. These might be based, for example, on Defenders of Wildlife's 1995 assessment of endangered ecosystems, or WWF's High Conservation Value Forests. Use information from Project 1 and other sources as appropriate to model the relation between patterns of forested land use (current and potential) and probability of maintaining/ restoring regional diversity. Forested land use includes reserves and their connecting corridors, intensive management, and various forms of ecosystem-centered management. Cooperative efforts among landowners, agencies, and NGOs are encouraged. Note, maintaining diversity refers to the functional viability of a species and not just the population viability. They are not necessarily the same. Species with key functional roles include those important in the food chain (eg truffles, lichens, Lepidoptera), in regulating pests (eg spiders, birds, parasitoids), and in cycling nutrients (the entire soil foodweb).

Basically we were trawling to see what we might hook. A number of proposals came in, most quite good, none touching all the bases outlined above. Because of budget constraints, only two were funded. Within a couple of years we should find out how successful we have been in our first round goal of stimulating synthesis and outreach. Given the quality of research teams funded by the Commission, I expect we will be successful. It will take longer to tell whether we have actually helped changed practices on the ground, and longer yet to tell whether changed practices actually translate to enhanced conservation.

Strategies at the landscape scale

The set of idealised landscapes shown in figure 6 has been useful for me in thinking about landscape designs that balance economics and environment.

The scale is purposely undefined and, in the fractal spirit that nature seems attracted to, one could view these at scales ranging from region to site (eg at the site level the reserves could be group retention patches). The common element to all is a network of linked reserves: islands and bridges within a sea of managed landscape.[12] The managed landscape consists of three options: intensive fibre management, approaches that attempt to combine economically feasible productivity with stand-level habitat conservation, and approaches that yield some fibre but emphasise habitat. Moving from figure 6a to 6c yields increasing habitat protection with, at least in theory, decreasing fibre production, although the latter hypothesis has never been adequately tested (at least in North America) and we really don't know the costs of habitat protection in terms of long-term fibre production.

A basic point to be made in the landscapes of figure 6 is that while soils and water must be protected on every piece of ground, habitat probably does not (luckily, since we've already converted a great deal of habitat to strictly human uses). But if not on every piece of ground, the evidence as I see it argues that habitat must be protected on a substantial portion—perhaps 50 per cent or more to be successful in the long run, but that is a hypothesis to be tested in specific places. It follows that I'd rate the chances of diversity being maintained by the approach of figure 6a pretty low, especially if it dominated the regional landscape. The approach shown in figure

---

12 The figures draw from Harris' Multiple-Use Modules, Harris 1984.

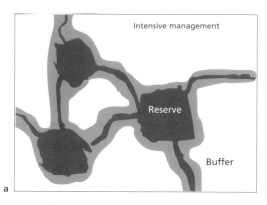

a

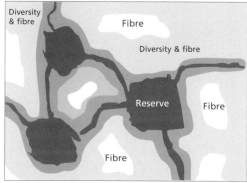

b

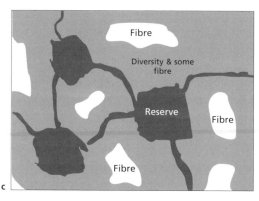

c

Figure 6 Various levels of multiple-use landscapes. The protection of biological diversity increases from (a) through (c), while commodity production decreases. The stability of commodity production could increase from (a) through (c), however, depending on how closely stability is tied to biodiversity. Solid: reserves and connecting corridors. White: focus on fibre (eg intensively managed plantation). Stipled: combined focus, with darker stipling indicating heavier emphasis on biodiversity. (Adapted from: Perry 1994)

6b is much like that adopted by FEMAT on federal lands in the range of the northern spotted owl: retention harvests in the matrix maintain basic structural elements (hence at least some at-risk habitats) and provide movement routes not confined to corridors (which may or may not be effective); reserves are located where needed rather than areas with low timber productivity. In figure 6c, habitat protection is stepped up a notch. Note, the habitat I'm talking about here is that lost in forestry regimes focused exclusively on fibre production, some of which might be protected under modified silvicultural regimes. An example relevant to many forest types, including mountain ash, is green-tree retention to provide a continuing source of cavity trees and large dead wood (which has important ecological functions) on managed lands.

In each of the cases shown in this figure, questions can be asked about the relationship between landscape structure and processes such as hydrology and the spread of disturbances. For example, in the fire-prone environment of the mountain ash forests, what are the implications of different landscape patterns for risk of spreading crown fires? Note that none of the scenarios, either at the landscape or site level, should be viewed as fixed ways of doing things, but rather as arbitrary points in a continuum of management options, their value not so much in providing fixed options but in stimulating thought and discussion that may open doors to new approaches.

The productivity (fibre) costs of nontraditional approaches will almost certainly vary depending on forest type, markets, etc, but it is important to be wary of overly negative assumptions. In the US at least, there is virtually no long-term, documented experience with alternative approaches, so we really don't know what the productivity trade-offs are. Over the past 10 years a number of long-term experimental comparisons have been installed in the US and Canada, but forest dynamics play out over long times, and it's particularly risky to draw conclusions about yields at rotation age from early tree growth. In the NCSSF we hope to draw out interpretable comparative information from 'natural experiments'— those scattered landowners who have done things differently for along enough to provide reliable information.

It is important to be clear at the outset about what, specifically, is to be maintained and how success will be measured. In the

9

Forest biodiversity management— the Swedish model

Per Angelstam

In response to society's concern about the world's forests during the 1980s and 1990s, considerable attention has been drawn to deforestation, loss of species, and consequently the need for sustainable forest management, restoration and re-creation of important structures where naturally dynamic forests have been lost. When considering forest biodiversity, it is important to understand the extent to which existing land use practices harmonise with the natural dynamics of the ecological past of the different forest environments. The key assumption behind this 'natural disturbance' paradigm is that native species have evolved under natural disturbance conditions. Thus, the maintenance of sufficiently similar conditions offers the best insurance for maintaining the species of that ecosystem. This has also been expressed as a 'coarse-filter' approach by which the characteristic ecosystems and landscapes are maintained.

For practical implementation, both the range of natural disturbance regimes and the resulting forest environments to which species have adapted (the ecological dimension), as well as the range of different land management regimes that can be applied (the management dimension) must be reasonably well known. Consequently, there has been an increased focus on trying to understand the ecology of forest ecosystems. In North America the concepts of 'new forestry', 'forest ecosystem management', 'natural disturbance ecology', and 'ecological forestry' have been advocated as tools for sustaining ecological integrity, including the maintenance and restoration of biodiversity. Similar ideas have been put forward in Europe, where the natural dynamic of forests has been proposed as a source of inspiration for woodland and forest conservation management.[1] During the 1990s the disturbance regime concept became a widely accepted approach to argue for applying an increased range of silvicultural methods. Still, however, clearfelling with variable retention is the norm.

The current Swedish forest policy defines the biodiversity maintenance objective as that 'all naturally occurring species should maintain viable populations'. Given the country's very long history of forest use and management, this is an ambitious goal, and an indication of the difficulties in reaching it is given by the viability status of species. Currently, a total of 2101 forest species of the 58,000 species listed in Sweden are listed as endangered and vulnerable. Partly due to the long land use history, the proportion of protected areas in managed landscapes is only about 2 per cent of the productive forests. Consequently, in the absence of a well-developed system of protected areas, the main tool to achieve the biodiversity goal is proactive forest management and restoration.

During the 1990s, the combined efforts of habitat management and tree retention at different spatial scales of private landowners, large forest companies, and the state have become a 'Swedish model' for reaching the biodiversity goal of sustainable forestry. However, the extent to which these ambitions will be reflected as a reduction in the number of endangered species and future maintenance of viable populations of all naturally occurring species remains to be seen.

---

1 For example, Peterken's (1996) 'natural approach' in western Europe, Remmert's (1991) 'mosaic cycle concept' in central Europe, as well as northern Europe (Angelstam 1998; Larsson & Danell 2001; Korpilahti & Kuuluvainen, 2002).

Swedish landscapes

Biogeography

Latitude and altitude are two basic abiotic factors affecting biodiversity. Being latitudinally extended, between the 55th and 69th parallels, Sweden's vegetation period varies from less than100 days in the north to greater than 200 days in the south. The Marine Limit, that altitude under which fine sediments rich in nutrients were deposited in a now disappeared sea shortly after Sweden was totally glaciated about 10,000 years ago, and the distribution of lime-rich soils have a fundamental effect on both the natural potential vegetation, and forest loss due to agricultural development. Further, prevailing southwesterly winds and higher altitudes in the northwest than in the east produce distinct gradients in climate and natural disturbance regimes, but also in the distribution of anthropogenic pollution from Western Europe.

Through effects on soils, nutrient accessibility, and climate, both altitude and latitude have profoundly shaped the distribution of natural vegetation types. From south to north, the main potential natural Swedish vegetation types are:

- broad-leaved nemoral deciduous forest with *Fagus sylvatica*, *Quercus robur*, *Tilia cordata*, *Acer platanoides*, and *Fraxinus excelsior*

- a hemiboreal transition with mixed deciduous and coniferous forest

- a wide belt of boreal forest with *Pinus sylvestris*, *Picea abies*, *Betula* spp., and *Populus tremula*

- subalpine forest and alpine tree-less environments at higher altitudes in the northwest.

Human colonisation of Sweden closely followed the retreating ice shield. However, the anthropogenic transformation of the landscape was considerably slower. Up until the Medieval Period, Sweden was settled up to the border between hemiboreal and south boreal forest in the interior, and far north along the coast of the Baltic Sea in the east. Later, areas of local forest use were intensified. Starting about 150 years ago, large-scale logging was extended gradually into the interior of north Sweden. Consequently, the deciduous forest in the nemoral zone in the south has over 5000 years of habitat loss and land use change, while the boreal and subalpine forests in the north has a land use history of less than 200 years.

A short Swedish history of nature conservation

Swedish forest conservation is young. Although large national parks in the mountains were established about a century ago, forest conservation areas became an important issue only in the 1970s. The Nature Conservation Law was established in 1964 and the Environmental Protection Agency was founded in 1967. The driving forces for the maintenance of forest biodiversity started with protests against large clearcuts and use of herbicides to remove deciduous trees in the early 1970s. In the government's clearcut investigation from 1974 the need for considerations of species was suggested and amended to the forestry law in 1979. During the 1980s, the National Board of Forestry carried out several education campaigns, which culminated in 'Richer forest'.

The long history of intensive forest and land management has resulted in a heavy human footprint on the Swedish forest. As an example, the amount of different forest components such as dead wood, large/old trees, deciduous and old forest have declined by 90–98 per cent compared with what is found in naturally dynamic forests. However, the emerging knowledge about critical thresholds

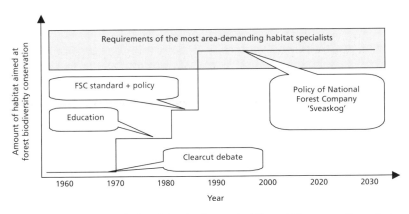

Figure 7 Schematic description of the development of the ambitions regarding nature conservation considerations (sum of protected areas, management and habitat restoration/re-creation) in Sweden during the past 30 years

for species at the level of individuals and populations makes it possible to start relating the short-term political goals and management achievements to the long-term vision of maintaining viable populations of all naturally occurring species (see figure 7). Consequently, it can be concluded that the protection of the very last remnants of forests of high conservation value, habitat restoration in the managed landscape, as well as habitat re-creation are all important management strategies for maintaining viable populations of all naturally occurring species. The 'Swedish model' can thus be seen as a grand attempt of forest biodiversity rehabilitation.

Forest protection and management

At present, the amount of forests that have been set aside in regional reserves and parks varies widely among regions. Although 44 per cent of the subalpine coniferous forests are protected, only about 2 per cent have been set aside for biodiversity maintenance purposes in the rest of the country.

Hence, even within a small country like Sweden, the variation in the degree of human transformation is large. In southern Sweden, although the proportion of natural forest was already seriously reduced in the nemoral forest zone by 1000 years BP, a large proportion of the landscape was covered by wooded pastures and meadows with large trees, providing refuge for forest-living species for several centuries. Today, therefore, a large part of many kinds of forest biodiversity in southern Sweden is connected with small remnants of the ancient cultural woodland landscape. In northern Sweden, by contrast, changes in the landscape are relatively recent and in the boreal forest regions which are regionally or locally remote, or which have a low potential for forest production, one can still find near-natural remnants, but not entire naturally dynamic landscapes any more. Inventories of such remnants, called Woodland Key Habitats, show than about 4 per cent of high conservation value forests which have note been protected still exist.

As a result of the biological and historic differences among regions, a variety of biodiversity management models of landscapes need to be developed. Above the Marine Limit in the upland north, imitating the dynamics of the natural landscape is much more feasible than in the lowland south. There the main solution is to maintain the small remnants of near-natural or cultural habitats left after a long history of land use. In southwest Sweden, the challenge of sustaining ecosystems affected by air-borne nitrogen pollution acid rain are additional and serious problems. Consequently, the Swedish experience is of general interest for those involved in forest biodiversity maintenance. This is true both where the forest history is complex and forest ecosystems need restoration to maintain their biodiversity, and where landscapes are still dominated by natural forest ecosystems and forests need to be managed and used in ways that do not threaten the existing biodiversity.

Biodiversity management tools across spatial scales

Considerations at each forestry operation
The first rules for nature conservation in the managed forest landscape from the 1950s included retention of trees and small remnants on sites with poor forest production, riparian forests and field margins, and deciduous trees near settlements, but only on state-owned land. By the 1970s, scientific arguments for nature considerations became stronger. At the end of the 1980s, 2 to 3 per cent of the total harvested volume of merchantable wood was left after clearfelling. A few years later, the level of retention across spatial scales from trees to landscapes had increased to 10 to 15 per cent of the area, but less by volume. The main reason was a massive education campaign by forestry staff directed at authorities, corporate companies, and private forest owners. Later a market-driven pressure speeded up this process somewhat.

Adaptation of management methods
Following the tradition of Tore Arnborg from the 1940s, the ecologist Jan-Erik Lundmark successfully advocated site-adapted forest management, whereby the local site and regional climate determined methods and selection of tree species and regeneration methods. Similarly, the habitat requirements of different species other than trees can to some extent be translated to the type and amount of forest in the landscapes. Because the location, composition, and structure of forest types of different disturbance regimes are determined by local and regional site differences, biodiversity management has also become more adapted to the site type.

148

In boreal Sweden, the development of different practical forest management regimes promoting biodiversity has mainly been based on the natural distribution of three disturbance regimes found in natural fire-driven dynamic of the boreal forest. Selective cutting systems and voluntary protection are promoted in site types and forest systems with natural *Picea abies* gap-phase dynamics, whereas clearcuts with variable tree retention are considered to be ecologically proper for sites with fire-adapted communities on mesic sites where stand-replacing fires are common. Finally, on dry sites where low-intensity fires have created multilayered cohorts of *Pinus sylvestris*, clearcutting with repeated retention of both live and dead trees in successive management steps, partly in combination with prescribed burning, is recommended[2]. While these differences in dynamic among forests on different site types are generally acknowledged, in reality there is a large discrepancy between what should be done and what is done in practise. Forest management is still largely dominated by clearcutting using variable retention (98 per cent of the harvested area) even if the mesic sites where this can be considered appropriate amount to about 70 per cent. Ideally, in the remaining 30 per cent with dry and wet sites, where disturbance regimes other than succession would prevail naturally, alternative silvicultural methods should be attempted.

In southern Sweden (nemoral and hemiboreal forest), some anthropogenic disturbance regimes, such as those found in the old cultural landscape, also retained many aspects of natural ecosystems and the associated forest species. Cultural disturbances such as low intensity forestry and farming resulted in a high species richness being maintained, and even increased compared with the previously forested habitats. This represents a complex form of endangered landscape using large areas to extract relatively few resources, little external input of energy and nutrients, and multiple use of several landscape components resulting in high species richness. However, as forestry and farming were intensified, biodiversity was reduced through loss of structural diversity from both forest and farmland. In order to maintain biodiversity in cultural landscapes, a practical, but expensive, solution would be to mimic those cultural disturbance regimes.

---

2 Fries et al 1997; Angelstam 1998

Landscape planning

In around 1990, a few scientists and forest managers began a cooperative effort to develop practical tools for managing biodiversity at the landscape level. They attempted several approaches, and their application shows strong connections to the biological and historical complexity of the region, as well as to the ownership pattern. For northern upland Sweden with boreal forest, where the land-use history is relatively short and land is owned mainly by large companies, the goal of planning is to imitate the natural disturbance dynamics of the different forest ecosystems. Given the site type, forestland is stratified with respect to different disturbance regimes with the potential wildfire dynamics as a guide. This ecological landscape planning is complemented with landscape ecological planning whereby remnants of the different types of forest dynamics are set aside with the long-term goal of maintaining sufficient habitat connectivity. In contrast to other parts of Europe, landscape design based on aesthetic considerations has not been explicitly used in forestry.

In Sweden, several large landowners have developed models for Ecological Landscape Planning (ELP), including multiple goals and spatio-temporal scales. ELP can be defined as a planning tool to systematically bring about a decrease in the differences in the amounts of different habitat properties between past primeval or little impacted forest landscapes on the one hand, and present managed landscapes on the other. These habitat properties have been defined by species which are listed as endangered, and by some processes which have disappeared or changed as a consequence of the development of intensive forest management for timber and pulp only. The following eight activities have been defined in the landscape planning process:

1   Stratification of forests into different biogeographic regions with their characteristic past disturbance dynamics, to which the fauna and flora have evolved, and into regions with different land use histories.

2   Landscape analysis to estimate the physiographic and historical potential of a particular landscape to host different habitats and properties. In reality, landscapes of 5000 to 50,000 hectares have been analysed, and even less

if the complexity of the area is high and the number of landowners is large. The range should also be determined by the size of an area that could hold viable populations of different species.

3 Description of the present composition, structure, and processes of the selected landscape. This includes inventories of so called woodland key-biotopes, wet forests, and red-listed species.

4 Analysis of components missing or insufficiently represented in the actual landscape to maintain or restore biodiversity.

5 Formulation of quantitative goals for each property and scale.[3] In short, the landscape analysis is the basis for estimating the composition, structure, and processes in the original landscape. Knowledge about ecological thresholds is applied to these figures.

6 Choice of a strategy for how to act in practical management. The strategy is mainly related to the type of ownership. In the north of Sweden a few decision-makers in large companies over 70 per cent of the landscape, whereas in the south of Sweden there are 10–1000 decision-makers in a landscape.

7 Implementation of forest management. The Swedish system defines different management regimes, ranging from no management to intensive management with nature considerations.

8 Control by monitoring to allow short term steering to attain long-term goals. To be effective, monitoring of biological systems must have a sound scientific basis, be diagnostic and help understand the system, allow assessment of the stated policy objectives and, finally, include feedback to the policy process and/or management to enable mid-course corrections.

Under continuous development, landscape planning is expected to become more complex as more ecological knowledge becomes

---

3 This step is the most difficult, and is treated in detail in Angelstam & Andersson 2001.

available and as more goals (eg social, recreational, and tourism-related) are added. In practical forestry, the amount of work invested by foresters in ELP most probably will need to be increased, along with the use of relevant basic information about different landscape components. With an increasing number of factors to be considered, the use of new types of habitat maps, including geographic information systems and various decision-support systems, is crucial. There is hence a need to complement the strategic part of ELPs with spatially explicit tactical planning to set cost-efficient priorities for conservation, management and restoration for different elements of biodiversity at different spatial scales ranging from trees and stand to landscapes and regions.

However, if these aspects are not dealt with adequately, ELP is of little value and could even be a way through which politically negotiated poor biodiversity management methods and conservation targets could be implemented in the short term at the expense of long-term success in biodiversity management. This is particularly crucial in the contacts between the West and the East in northern Europe.

A chronology of forest certification in Sweden

From 1990, forest certification was initiated by environmental NGOs such as the Swedish Nature Conservation Society and WWF-Sweden to promote the development of sustainable forest management as well as to provide a better market for products that meet certain standards. In an analysis of the development of forest certification programs in Indonesia, Canada and Sweden, Elliott and Schlaepfer[4] concluded that certification can be best understood as a policy instrument which promotes and facilitates learning among stakeholders, both during the development and implementation of standards. Hence certification provides direct incentives for improved forest management. Moreover, the consensus-building among actors such as NGOs, forest owners, indigenous people, and government—who have traditionally been in conflict with each other—can be significant. Being initiated by NGOs, forest certification has in some countries been seen as a threat to

---

4 Elliott & Schlaepfer 2001.

government forest departments or the forest industry.

Currently, forest certification in Sweden is largely based on two ideas. The first approach was established by the Forest Stewardship Council (FSC), which defines three goals for forest management: sustaining economic wood production, forest ecosystem function, and social development. Naturally, the success of this complex task depends on the ownership pattern and the skill, interest, and economic potential of the owner, as well as of the condition of the forest system itself. In Sweden, the forest industry has been positive to FSC while private landowners' associations and the National Board of Forestry have been negative or indifferent.

The process of developing a national FSC standard started in 1993, with a broad representation of different stakeholders from the forest industry, the forest-owners' association, WWF-Sweden, and the Swedish Society for Nature Conservation (SSNC). The first proposal of a national standard was presented in May 1995. However, at the time the different forest stakeholders generally hesitated to agree, and the preliminary criteria, presented by WWF and SSNC without consent from other stakeholders, contained mainly ecological aspects. During the following year, a certification company made the first field tests of the May 1995 criteria. Simultaneously the forest industry considered alternatives for achieving the market's acceptance of the current ways of managing forests, and also sought to determine the level of market pressure for certification. At the time, the forest-owners' association did not approve of the idea of FSC certification. Two important reasons were that they could not accept the limitations with respect to the forest management in the remaining near-natural subalpine forest, and the consideration to the indigenous Sami people for reindeer winter feeding areas in northern Sweden. However, there was a small group of individuals in different stakeholder groups who refused to give up the idea of forest certification according to FSC model.

In 1996, it became evident that large forest enterprises did respond to market pressure. Although the revision of the May 1995 FSC standard was still in progress in a new round of discussions, one of the most internationally active companies (STORA Forest) initiated a voluntary FSC certification of one forest management district. Although the assessment indicated some gaps in current

management practices, it showed that forest certification was prac-
tically feasible. In 1997, several large companies embarked on the
certification process, and Korsnäs Forest became the first FSC and
ISO 14001 certified forest company in the world. In January 1998,
the FSC approved the Swedish national standard, and it became the
first voluntarily negotiated national standard for forest certification.
Even so, most private landowners remained negative about FSC
certification. In spring 1998, a group of private landowners with
relatively large land holdings was assessed under the FSC standard
and passed. Yet the forest-owners' association representing small
landowners has not embarked widely on FSC certification. In
practice, however, some of the guidelines recommended to their
members (eg the forest-owners' association Södra in southern
Sweden) are similar to those supported in the FSC standard.

In Sweden, a clear latitudinal gradient exists in the proportion
of privately owned forest from south (about 80 per cent) to north
(about 30 per cent). In addition, the pattern in the size of holdings
of the different ownership categories is clear. For corporate
companies, 98 per cent of the land holdings are over 400 hectares in
area. By contrast, for privately owned forest, only 11 per cent are

Table 2 Chronology of forest certification in Sweden[5]

Year	Description of what happened
1992	Consultations by FSC in Sweden begin
1994	WWF-Sweden forms a reference group to provide advice on the development of a Swedish forest certification standard
1995	WWF and the Swedish Society for Nature Conservation present criteria for nature conservation in forests
1996	First certification of large company (STORA) (Rhubes et al. 1996) Official FSC working group is created in Sweden
1997	First FSC and ISO 14001 certified forest company (Korsnäs AB); (Brunberg and Johansson 1998)
1998	First association of large private landowners (Skogssällskapet certified)
1999	First association of small private landowners (LRF) embark on FSC
1999	Process to establish a Swedish PEFC scheme starts
2001	Attempt of harmonise the standards of FSC and PEFC fails
2002	First PEFC standard
2003	Revised FSC standard is expected to appear

5 Modified after Elliott 1999.

over 400 hectares. The interest in FSC certification is clearly related to the regional differences in the pattern of ownership, which in turn is clearly related to the sensitivity of different ownership types to international market pressure. As a consequence, the large forest companies embarked on forest certification first while smaller companies followed later (see table 2). This is probably due to the fact that only sufficiently large forest owners have direct contact with the market and can perceive a market pressure. For small private landowners, the lack of interest is not unexpected as wood can be harvested and sold only a few times per decade.

In 2003 a new revised Swedish FSC standard should be presented. Simultaneously, an international discussion is ongoing about how regional and national standards should be harmonised. This harmonisation can be made at two levels: the variables included, and the targets to be reached.

The second approach was that of Pan-European Forest Certification (PEFC). The aim of PEFC is to offer long-term assurance that the certified wood comes from sustainable managed forests in which sound conservation is practised. It was founded in 1999 and is based on the Lisbon-declarations adopted in the Third Ministerial Conference on the protection of forest in Europe (Liaison Unit in Lisbon 1998). In late autumn 1999, the Swedish forest owner's associations and the private independent sawmills of Sweden initiated a process to establish a Swedish PEFC scheme. In 2000, the first national standards were approved, those of Sweden, Finland, Norway, Germany, and Austria.

There are, however, important differences between FSC and PEFC. In fact, in 2001 an attempt to harmonise the Swedish FSC and PEFC standards failed. Of 18 observed differences PEFC has so far only agreed to change three. There is still a considerably unwillingness among private landowners to accept that other stakeholders should have a say about how private forests should be managed. Moreover, the Sami issue has not yet been solved with PEFC.

Another difference between FSC and PEFC is that the forest-owners' associations have sometimes allowed forest owners to become certified free of charge, while FSC certification requires external reviewers.

To conclude, forest certification has been an efficient tool for mutual learning among a wide range of stakeholders in the forest

sector. It is, however, vital to stress the difference between certification of environmental management audit systems such as
ISO 14001 or EMAS on the one hand, and environmental standards,
such as FSC and PEFC, on the other. While the former implies that
a framework ensures that environmental issues can be dealt with in
a consistent way, the latter sets the goals that should be reached.
Another interesting benefit of certification is that while public
policies usually change slowly over decades, the private policies of
forest product companies and retailers can adapt more rapidly to
changing circumstance.

As to the targets, every certification standard is a compromise
of different interest groups. Consequently, concerning the amount
of suitable habitat for different species, different countries are
located at a different stage, and standards can be expected to be
more similar to the current status of the forests than to the
requirements of viable populations of species. However, it should
be noted that the ecological knowledge of the quantitative
requirements of species at the scale of trees, stands, and landscapes
is only one of several factors to be included in this negotiation
process.

In principle there are several ways of combining variable
retention, ELP and protected areas. With a long land-use history
and today's values concerning biodiversity, Sweden thus suffers an
environmental debt, which requires restoration and re-creation
from beneath in relation to the emerging insights about critical
habitat thresholds. The 'Swedish model' is therefore the only
solution in Sweden.

With an international perspective, however, it is vital to
separate the discussion about what regions and countries like
Sweden are limited to, as opposed to the choice of different combinations of forest management, protected areas, and habitat re-
creation, which other countries and regions have the flexibility to
choose. In general, the degrees of freedom are considerably larger
when the length of the land-use history is short. This is particularly
evident in Europe where forestry in Eastern Europe is rapidly
becoming more intense. We have to realise that the goals that are
discussed in relation to mechanisms, such as the Swedish FSC-
standard, are political and not ecological. If we do not reach this
point, there is a considerable risk that the still much more near-

natural forests in Eastern Europe will be transformed and biodiversity lost in a very near future, as has happened in Sweden and other countries in Western Europe in the past.

Analysis of gaps in forest protection

While it indeed appears possible to maintain viable populations of many of the naturally occurring species in combination with timber and pulpwood production in several of the forest types found in Sweden, this is not true for all species in all forest types. The reasons are usually lack of sufficient amounts of specific forest components (eg certain types of dead wood, deciduous trees, old trees) and forest types (eg wet old-growth spruce forest). Moreover, for some species large areas of unmanaged forest are required. Hence forest reserves are an important component in the efforts to maintain forest biodiversity.

A critical issue is what forest reserves are needed. In 2001, an estimate of the need for set-aside of forest reserves of Sweden's main forest types to maintain long-term viable populations of all occurring forest species, both in typical forests and in woodland in cultural landscapes, was presented to the Ministry of Environment.[6] The starting point was to deduce the potential occurrence of 14 different forest types from the database of the National Forest Survey. Boreal forest types were deduced from the present site type distribution while for nemoral and hemiboreal forest types the occurrence of broad-leaved tree species as well as cultural landscape woodland was also used.

Gap analysis can be defined as the disproportionate scarcity of certain ecological features in a management unit, relative to the representation to a larger region surrounding the management unit. In order to estimate the amount of the different forest types that need forest reserve status, three different aspects were considered. First, the disturbance regime of each forest type was summarised and analysed to understand the extent to which different types of external and internal disturbances maintain their characteristic dynamics. This was then compared with the potential achievements in the appearing new more nature-friendly forest

---

6 Angelstam & Andersson 2001.

management regimes using the expert opinions of practitioners to determine to what extent management can emulate the natural disturbance regime. If emulation appeared possible, the need for protection was reduced. Finally, the emerging knowledge on the critical habitat loss threshold values of 70–90 per cent for metapopulation persistence was used to estimate what proportion of each forest type is required for the long-term survival of species found in that forest type.

The proposed forest reserve needs therefore represented only forest types which cannot be sustained using forest management methods employed within the framework of the Swedish model for biodiversity maintenance. The results of the analysis for Sweden were divided into four biogeographic regions: nemoral, boreo-nemoral, south boreal and north boreal. Due to differences in the possibility of emulating important disturbance regimes and the mix of different Swedish forest types in different regions, the estimated long-term requirement for maintenance of all forest species ranged from 9 per cent of the forestland in the north boreal to 16 per cent in the nemoral region. Using the data of the national forest survey it was then estimated that the remaining amount of unprotected forests with old-growth characteristics was 3 per cent or 700 000 hectares, ie very close to what was later estimated to exist in the Woodland Key Habitat inventory.

The gap between the long-term goal for forest protection and available forests to protect suggests that it is urgent that the amount of protected forest is increased to encompass the remaining very few forests with a high conservation value, and that forest protection alone is insufficient to reach the biodiversity maintenance goals. The long history of land-use change over the past 150 years also calls for large-scale forest ecosystem restoration. Due to regional differences in the extent of past land-use changes, this need of restoration increased from the north boreal forest (3 per cent) to the nemoral forest (11 per cent). The clear trend in restoration needs is also evident from the forest situation in Europe. Although about 20 per cent of the boreal forest in Europe is not or is little affected by human disturbance, the corresponding figures for hemiboreal and nemoral forests are 2 and 0.2 per cent, respectively.[7] However,

---

7 Hannah et al 1995.

habitat restoration, let alone forest re-creation, is hardly on the agenda in today's Swedish forest management at all.

There are also important functional aspects of biodiversity which need to be addressed. About a quarter of the forest soils are so affected by acid rain that base cations are lost from the ecosystem by leaching. Nitrogen pollution is changing the species composition of fungal communities. Finally, in the absence of functional populations of large carnivores in most of Sweden, large populations of deer and moose alter the tree species composition.

Future challenges

During the 1990s there was a strong international trend in forest management trend towards having to satisfy several other objectives in addition to wood production. In Sweden, the new main challenge became to maintain forest biodiversity. As a consequence, forestry and society are currently making increasing investments in new types of management, which include retention of timber, as well as set-asides of forest in reserves. However, the relative contribution of these activities to the maintenance of biodiversity, as well as how large these investments need to be under different circumstances, is not fully understood. The following questions concerning the functionality of the Swedish model for sustainable forestry are therefore commonly raised:[8]

- How much forest do we need to set aside in forest reserves?
- How much timber do we need to leave in the matrix surrounding the forest reserves?
- What can we do in different management steps to restore/re-create important features that are required to maintain biodiversity?

However, given the fact there are more than 2000 red-listed forest species in Sweden, nothing suggests that the efforts are too large.

---

8 For example, Larsson & Danell 2001.

The need for syntheses and communication

The Swedish experience during the 1990s is that attitudes can change rapidly in favour of completely new management goals. However, ensuring long-term success—that all naturally occurring species survive in the long term—is a grand commitment. To bring about the success of new initiatives and eventually reach the long-term goal of maintaining viable populations of forest species in Europe's landscapes, it is essential that the transfer of experiences between scientists and practical management is satisfactory. Creating operational goals and to demonstrate progress is difficult, and there are several challenges:

- Visions: The first challenge is to have clear definition of the goal. As the biodiversity concept originated from the concern about the loss of species, the obvious short-term goal is that negative trends among authentic species should be removed. From this follows that the long-term goal, or vision, is that the conditions for populations will be such that species have viable populations.

- Tools: The second challenge is to define the tools that allow us to reach both the short-term and long-term goals. These tools range from creating awareness and ensuring continuous capacity building, to the actual planning and management techniques.

- Benchmarks: The third challenge is to know when we have reached the goal. This includes both the technique of monitoring success in reaching short-term goals and that quantitative benchmarks can be documented and targets derived.

- Communication: Finally, both the monitoring results and the targets must be communicated and understood by the different actors in forest management.

Because of the European Community, large parts of Europe are united in overall policy and general ideas on biodiversity. By contrast, the status of and threats to biodiversity, as well as the ecology and management of forest environments in the various regions, are highly variable. This implies that there is usually no clear single vision, nor are there regionally available simple tools for how to choose from the biodiversity management package containing set-aside in reserves, considerations in the managed forest, or forest restoration. Neither is there a unified system for assessing to what extent the biodiversity maintenance goal has been reached in a particular region.

Because the debate on how to realise the goal of maintaining biodiversity was addressed early on in Sweden, there are experiences, both positive and negative, that should be communicated to other regions in progress. For scientists, the challenge is to collect more information about critical thresholds regarding the amount of habitat required for different species groups at different spatial scales to maintain viable populations. In addition, there is a need to develop, in cooperation with forest managers, knowledge and tools to aid the spatially explicit practical planning for sufficient habitat connectivity when quantitative goals for different habitats have been formulated.

The Swedish model and staff reduction

At present, stand considerations, new management methods, ecological landscape planning within a market-driven forest certification, and set-aside of forest reserves all contribute to biodiversity maintenance in a 'Swedish model' of biodiversity management where all land-ownership categories are contributing. The challenge is to sustain these efforts in the long term.

The strong reduction in habitat structures such as dead wood suggests that long-term maintenance of viable populations of many species require effective habitat restoration. Hence, it is essential that the new proactive site-adapted forest management and the considerations in all forest management steps can be sustained for many decades. In regions with a long history of forest management and other kinds of land use, habitat restoration means establishing new stands by wise planning and silvicultural treatment of young forest stands. However, an intense rationalisation to cut costs over the past three decades has led to a strong reduction in staff both in forestry and other land management organisations. For the forestry sector, the Swedish National Audit Office has noted that although the forest policy is very ambitious, the law itself is very weak, and is insufficient to reach the goal of the policy. Due to this paradox, the main instrument is to provide advice and recommendations to forest owners. However, within the National Board of Forestry's field organisation the number of staff members is being reduced; moreover, this organisation has very few staff members with a biological education.

Regional planning of green infrastructures

The increased state funding to acquire the remnants of near-natural forest and set them aside as reserves provides a major challenge to buy land in a strategic way that maximises the effect on maintenance of viable populations. A major limitation is the poor cover of spatially explicit land-cover data describing different forest types with a sufficient thematic resolution across the whole landscape. In fact only for one of 22 counties is there a complete digital land-cover classification, and in three other counties such information is in progress. Note, however, that this information is not thematically ideal for biodiversity management purposes. It is also important to create demonstration areas to illustrate problems and possibilities of how protection, sustainable management, and restoration of forest types can act in concert under different conditions. Within about five years a working model for strategic planning of where the efforts to set aside forest land both with present and future conservation values needs to be in place. It is also important that other actors, such as the municipal administrations, become aware of the need to build green infrastructures for maintaining forest biodiversity.

Monitoring systems

A critical question is how progress should be gauged. One of the major problems with biodiversity management is the required long-time and large spatial perspective. While the economic and political time scale rarely exceeds 3 to 5 years, scientific knowledge can be measured by the decade, and the forest itself has a lifespan of at least an order of magnitude of 100 years or more. There will, therefore, always be a risk that long-term benefits for biodiversity maintenance are neglected due to immediate or short-term economic or political benefits.

It is therefore crucial to evaluate the relative biological importance of different biodiversity management components in both the short and long term. Still, no system to cover all components of biodiversity across spatial scales is at hand, either at the national level or at the scale of forest management units. It is also crucial to ensure that the results of monitoring and assessment systems are understood among different forest stakeholders. The following points describe a tentative logic by which a biodiversity assessment system can be developed in an iterative process involving both science and practice.

- Managers and scientists identify present-day problems with the implementation of forest biodiversity conservation policies in practical case studies.

- Based on the experiences from the mutual learning, a state-of-the-art practical assessment system is derived to cover the components of forest biodiversity and management scales.

- Applying and refining the biodiversity assessment system in a range of test sites representing the ecological and economic spectrum in the region/country with respect to:

  a. the status of biodiversity in relation to the reference/benchmark

  b. the possibilities to apply iterative feed-back between science and land use practices

- Pedagogic and practical management tools are developed to reconcile the maintenance of forest biodiversity and economic activity.

- The experiences from science and practice are communicated to policy-makers, academic institutions and agencies.

Critical habitat loss thresholds

In order to assess the status of biodiversity, the results of the monitoring process must be compared with some kind of benchmark. Only in this way do we know when the long-term maintenance of viable populations and systems function has been achieved. A question of imperative importance is the critical loss of habitat for different forest types and spatial scales that causes populations of different species to go extinct. To answer this question, the following activities, which aim at linking management regimes with biodiversity requirements into assessment systems, could be envisioned.

- identify the range of management tools and regimes being used to maintain biodiversity
- stratify the forest into different disturbance regimes as this is what species have evolved with. This will include an inventory of the geographical distribution of different disturbance regimes with the aim to produce a stratification of the forest into groups with characteristic mixes of different disturbance regimes.
- describe the historical spread of different kinds of anthropogenic impacts on the boreal forest (ie local use, exploitation, intensive management, and 'new forestry'). This would allow us to identify replicates of forest impact in regions with different disturbance regimes.
- identify response variables that are affected by habitat loss. One approach is to identify 'meta indicator species' grouped by both habitat requirement needs and biological life history traits. Ideally, such groups of species with similar habitat requirements and similar life history traits, and which are sufficiently charismatic and/or interesting, can be used as pedagogic 'messengers' to communicate the results to managers and other stakeholders. In this way analyses can be made both using traditional empirical data and by exploring the effects of different levels of habitat loss on species with certain combinations of life history traits using modelling and simulation.
- identify the monitoring and assessment 'currencies' that are both relevant and possible to use, to communicate the status of habitat loss across different spatial scales in landscapes with the different ownership pattern and management regimes that are found (eg tenure system, corporate companies owning land, private landowners)

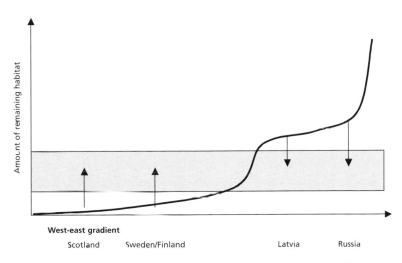

Figure 8 Illustration of the problem of trying to strike the balance between forest use and nature conservation in European boreal forests. The black line illustrates the range of remaining amounts of authentic habitat in the west–east gradient in northern Europe. The coloured interval represents the tentative range of threshold values to be exceeded for the maintenance of viable populations of forest specialists with large area requirements. Finally, the arrows represent the need for restoration in the west, and the need for pro-active planning in the east.

The European boreal forest as a time machine

The European boreal forest, extending from Scotland to the Ural Mountains, provides a unique resource for the gradual development of sustainable forest ecosystem management. The reason is the steep gradient in land-use history whereby the gradual exploitation and intensive management of boreal forest resources has spread like a tidal wave from areas of high demand to more and more remote regions. This 'time machine' allows us to understand the effects of the human footprint on the boreal environment.

At first glance, this area may seem huge and without threats to its biodiversity as a whole. However, recent studies show that even in European Russia a surprisingly small area (13 per cent) of what could be called intact natural forest landscapes remains today. Further west, such intact and productive areas do not exist at all. In fact, even if considering all the productive boreal in Sweden only about 5-6 per cent can be termed forests with high conservation value. In Scotland, most natural forests have been lost—only about 1 per cent is left. The maintenance of boreal forest biodiversity is therefore evidently a matter that concerns the European boreal forest as a whole. It is therefore important that both the west and the east share the relevant biological knowledge of importance for the long-term maintenance of boreal forest biodiversity (see figure 8).

- In the west, maintenance of biological diversity requires restoration of habitats to reach the threshold in the future.

- In the east, the development of forest management should not proceed so far that it reduces the amount of important forest components below this threshold

Acknowledgements

This paper is an extension of a report given at Oregon State University in 1998. My sincere thanks go to Bond and Barte Starker whose sincere interest in forest management and financial support to the Starker lecturers help to encourage a most stimulating exchange of knowledge and ideas. Big thanks to David Lindenmayer who took the initiative to organise a most stimulating meeting! Lena Dahl, Lena Gustafsson, Rosanna Mattingly, Erik Normark, Stig Larsson, Terry Rolfe and Ed Starkey provided clarifying comments on an earlier version of the manuscript.

10

Sustainable forest management in New Zealand

David Norton

Historically, New Zealand was almost totally forested, with only those areas above the alpine timberline, on recently disturbed sites, and on sites too wet to support woody vegetation lacking a forest cover. Even in the driest areas the historical evidence suggests that woody vegetation occurred, albeit forming open woodlands. However, tall forest dominated over most of New Zealand. As a result of 700 to 800 years of human settlement, by people of both Polynesian and European origin, the area of indigenous forest has been substantially reduced and now comprises about 24 per cent of the land area. In the last 100 years extensive areas of plantation forests (mainly radiata pine, *Pinus radiata*) have been established and now occupy about 1.8 million hectares (or about 7 per cent of the land area).

For most of the 20th century, the New Zealand Forest Service (NZFS) and its predecessors managed forests for a range of values including timber production, watershed protection, recreation, and biodiversity conservation. However, as a result of major government restructuring in the 1980s, private companies now manage almost all plantation forests, while the Department of Conservation (DOC) manages most indigenous forests for their conservation and recreational values alone. While not unique, the lessons that can be drawn from the development of sustainable forest management in New Zealand offer much to other countries.

A brief history of New Zealand forests and forestry

New Zealand's indigenous forests are structurally dominated by members of the Podocarpaceae and Fagaceae families, with the most important forest trees including rimu (*Dacrydium cupressinum*), totara (*Podocarpus totara*), kahikatea (*Dacrycarpus dacrydioides*), matai (*Prumnopitys taxifolia*), and the four species of beech (*Nothofagus*). Other important forest tree families include the Lauraceae (tawa, *Beilschmeidia tawa*), Cunoniaceae (kamhai, *Weinmannia racemosa*), Myrtaceae (southern and northern rata, *Metrosideros umbellata* and *M. robusta*, manuka, *Leptospermum scoparium*, and kanuka, *Kunzea ericoides*), Elaeocarpaceae (hinau, *Elaeocarpus dentatus*) and Auracareaceae (kauri, *Agathis australis*). New Zealand's indigenous forests are rich in understorey plants, especially pteridophytes and bryophytes. Major drivers of forest pattern include growing season length (as expressed by changes in latitude and altitude), rainfall, landforms, soil fertility, and disturbance.

The indigenous Maori people, who are thought to have first settled New Zealand in about 1200–1300AD, used a range of forest plant and animal resources. Although some of this use was sustainable, much was unsustainable including substantial forest clearance, mainly by fire, and the extinction of many animal species (eg moa). Total forest area is thought to have been reduced by about one-third at this time.

Although the first known European contact with New Zealand occurred in 1642, it was not until the 19th century that significant European exploitation of forest began. The initial exploitation was totally unsustainable, and in many areas rapid and almost complete loss of forest occurred through milling and burning. For example, on Great Barrier Island off the northeast coast of North Island, kauri logging started in 1794 and over 150 years resulted in the virtual elimination of kauri forest from all but the most inaccessible parts of this 28,500 hectare island. A similar pattern occurred on Banks Peninsula, eastern South Island, where forest cover was reduced from 70,000 hectares with first European settlement (1830s) to 800 hectares in the 1920s through a combination of milling and burning. Much of the forest loss in New Zealand had occurred by the start of the 20th century, especially in easily accessible lowland areas, although in some regions ongoing forest clearance for timber production and farming continued through into the second half of the 20th century. European settlement is thought to have reduced forest cover by another third of its original extent.

More recently, the removal of subsidies to farmers for clearing regenerating forest has resulted in substantial areas of marginal farmland being abandoned and now regenerating back towards indigenous forest. For the Banks Peninsula example, some 28,000 hectares of land that was under pasture in 1900 has reverted or is now reverting back towards indigenous forest. As a result, the total area of indigenous woody vegetation in many parts of New Zealand, albeit much of it still regenerating, is higher today than it has been in the last 100 years.

The two major legacies of forest clearance in New Zealand are the confinement of remaining forest to small often highly isolated remnants and the disproportionate loss of some forest types. The effects of both these processes are most intense in lowland areas, with forest types on the most fertile soils all but gone in some areas. For example, the Brunner Ecological District (58,704 hectares) on the west coast of South Island was predominantly forested when Europeans first settled this area in the late 19th century. Within this district about 60 per cent of the land area has remained relatively unmodified and still carries a predominantly indigenous vegetation cover. However, the distribution of this unmodified vegetation is not even. The forests, shrublands and grasslands of the uplands are almost all still present, as are most of the lowland hill forests (podocarp kamahi forests). In contrast, less than half of the rimu–miro (*Prumnopitys ferruginea*) forests of the infertile outwash gravels remain, while only 12 per cent of the kahikatea forests and associated wetlands of the fertile recent alluvial surfaces remain. The reason for the small remaining area of kahikatea forest is because the fertile alluvial surfaces on which they occur have been intensively developed for agriculture, especially dairy farming. Not only is the distribution of remaining unmodified vegetation uneven, but the tenure of these vegetation types is also uneven, with the almost all the upland vegetation in the public conservation estate (managed by DOC) while most of the kahikatea forests are under private ownership.

While the rate of forest clearance for agriculture declined through the 20th century, it was the ongoing unsustainable harvesting of indigenous forests for timber that generated the critical political debates that had far-reaching ramifications for forest management in New Zealand.[1] Concerns were being raised early in the 20th century by eminent New Zealand foresters and forest scientists about the destructive impacts of harvesting based on clearfelling and subsequent wildfires, and the overall rate of forest loss. But despite this, the exploitation of indigenous forests for timber continued relatively unabated until the 1950s. A number of experimental approaches aimed at harvesting these forests more sustainably were gradually introduced (eg strip-felling and various selection logging systems), but large areas of forest continued to be clearfelled generating much negative public reaction. By the early 1980s, harvesting based on coupe systems (0.2–2 hectares) that were replanted with nursery-raised rimu seedlings were being experimented with. However, the public profile of indigenous forestry had by now reached an all-time low and, despite the best intentions of some scientists and foresters involved, these developments were overtaken by political events that were reshaping environmental management in New Zealand.

Despite attempts by the NZFS to slow the rate of cut and implement sustainable management of indigenous forests, it was recognised in the 1920s that the indigenous forests would not be able to provide New Zealand's timber needs indefinitely. In fact, one assessment in 1924 suggested that 75 per cent of the indigenous forests would be gone within 80 years. The results of experimental planting trials with a range of exotic conifer species dating from the 1890s provided the basis for an ambitious exotic conifer plantation establishment program from 1925 to 1935 which saw more than 100,000 hectares planted over this time. This program was aimed at providing New Zealand with a sustainable timber supply independent of the indigenous forests. Further planting booms followed in the 1970s and 1980s, and again in the 1990s (see figure 9) to give a total plantation area of approximately 1.8 million hectares today.

---

1   Roche 1990

The predominant species in these plantations is radiata pine accounting for 89.8 per cent of the total plantation area, with Douglas fir (*Pseudotsuga menziesii*) the next most important species (5.3 per cent).

While the area of plantation forests and their contribution to the national economy was expanding rapidly, it was the ongoing debate over the management of indigenous forests—and especially a desire for greater accountability in the use of public funds—which precipitated the demise of the NZFS in 1987. A combination of lobbying from environmental non government organisations (NGOs) together with a strong push towards more transparent accountability of government agencies and privatisation of the government's commercial activities saw the NZFS disestablished and its functions split between conservation and production. Management of the majority of indigenous forests was transferred to DOC, which operates under legislation that specifically excludes commercial timber extraction,[2] while a much smaller area of production indigenous forest and the government's plantation forests were transferred to a state-owned trading company (New Zealand Forestry Corporation Ltd). Subsequently, the cutting rights to most of the plantation forests were sold to private companies, while the government retained its interest in the production indigenous forests and some small areas of plantations, mostly on the west coast of South Island, through another state-owned company (which eventually became Timberlands West Coast Ltd).

The legislative framework

The approach taken to forest management in New Zealand, as elsewhere, is strongly influenced by legislation, and in New Zealand the 1991 Resource Management Act (RMA) is critical. This act applies to all private land in New Zealand, including Maori land, and sets out the principles that local authorities (regional and

---

2   Section 30 of the Conservation Act 1987 states (subsection 3) that 'Except as provided in subsection (2) of this section, the Director-General shall not authorise any person to take any indigenous plant on or from a conservation area for the purpose, or with the intention, of deriving gain or reward, whether pecuniary or otherwise, from its wood.' Subsection 2 allows 'The Director-General [to] authorise any person to take on or from a conservation area any plant intended to be used for traditional Maori purposes.'

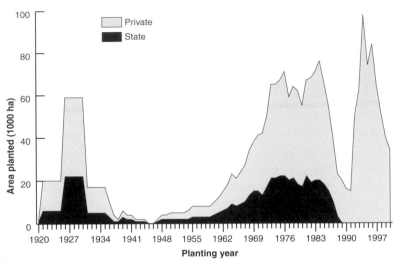

Figure 9 New land planted in plantation forests in New Zealand by the government (state) and by private companies (Ministry of Agriculture and Forestry statistics). Note that data from the 1920s and 1930s is from 5-year averages.

district councils) must take into account in determining what activities can occur in their area of jurisdiction. In terms of forestry (plantation and indigenous), the regional policy statements and district plans developed under RMA can prescribe the type of forestry that is acceptable, and the areas where it can and cannot occur (for example, because of high ecological values), and impose a wide range of conditions on activities such as harvesting. Within a plan or policy statement, all activities undertaken must be consistent with the purpose of the act (Section 5(1)) which is: 'to promote the sustainable management of natural and physical resources.'

Sustainable management is defined as (Section 5(2)):

'managing the use, development, and protection of natural and physical resources in a way, or at a rate, which enables people and communities to provide for their social, economic, and cultural wellbeing and for their health and safety while:

(a) Sustaining the potential of natural and physical resources (excluding minerals) to meet the reasonably foreseeable needs of future generations; and

(b) Safeguarding the life-supporting capacity of air, water, soil, and ecosystems; and

(c) Avoiding, remedying, or mitigating any adverse effects of activities on the environment.'

Furthermore, the RMA sets out some specific issues that local authorities must consider in administering the Act. In section 6 a number of matters of national importance are identified:

'In achieving the purpose of this Act, all persons exercising functions and powers under it, in relation to managing the use, development, and protection of natural and physical resources, shall recognise and provide for the following matters of national importance:

(a) The preservation of the natural character of the coastal environment (including the coastal marine area), wetlands, and lakes and rivers and their margins, and the protection of them from inappropriate subdivision, use, and development:

(b) the protection of outstanding natural features and landscapes from inappropriate subdivision, use, and development:

(c) The protection of areas of significant indigenous vegetation and significant habitats of indigenous fauna:

(d) The maintenance and enhancement of public access to and along the coastal marine area, lakes, and rivers:

(e) The relationship of Maori and their culture and traditions with their ancestral lands, water, sites, waahi tapu, and other taonga.'

In Section 7 a number of additional matters that local authorities must have regard for are identified:

'In achieving the purpose of this Act, all persons exercising functions and powers under it, in relation to managing the use, development, and protection of natural and physical resources, shall have particular regard to

(a) Kaitiakitanga:

(b)   The efficient use and development of natural and physical resources:

(c)   The maintenance and enhancement of amenity values:

(d)   Intrinsic values of ecosystems:

(e)   Recognition and protection of the heritage values of sites, buildings, places, or areas:

(f)   Maintenance and enhancement of the quality of the environment:

(g)   Any finite characteristics of natural and physical resources:

(h)   The protection of the habitat of trout and salmon.'

The RMA seeks to promote wise land management in dealing with the range of competing land management objectives, in particular to sustain ecosystem integrity. The act does not promote the separation of production and conservation, but rather promotes an approach that focuses on the sustainability of all aspects of ecosystems, including human needs and aspirations, within the same landscape. At the same time, however, it also recognises that there are a number of environmental 'bottom-lines' (the safe-guarding bit of Section 5, and the matters covered in Sections 6 and 7) that must be considered in assessing the suitability of a particular development proposal.

The RMA is not the only legislation that promotes integrated land management with respect to forests. The purpose of Part III of the Forests Act 1949, as amended in 1993, is (Section 67B): 'to promote the sustainable forest management of indigenous forest land.' The act defines sustainable management as (Section 2):

'the management of an area of forest land in a way that
maintains the ability of the forest growing on that land to
continue to provide a full range of products and amenities in
perpetuity while retaining the forest's natural values.'

The Forests Act implicitly recognises that indigenous forests managed for timber production as well as non-production values will not be the same as forests where no timber production is permitted such as those managed under the 1987 Conservation Act. However, this part of the Forests Act does not apply to exotic plantation forests.

The Forests Act provides for the conservation of indigenous biodiversity (natural forest values) through regulations governing sustainable forest management permits and plans (eg through limiting the number of trees harvested and harvest methods). The Act explicitly takes into account both the provision of the range of products and amenities that the forest can provide, and the retention of the forests natural values. Although natural values are not defined in the Act, they are taken to include the biological (flora and fauna) and physical (geology, soil, and water) features of a forest. Thus timber harvesting does not take precedence over natural values or vice versa.

Both pieces of legislation are similar, although the Forests Act has a more narrow focus, and both position indigenous biodiversity conservation as occurring with some sort of economic activity, not as mutually exclusive land uses. This is in sharp contrast to other conservation legislation (eg 1977 Reserves, 1980 National Parks and Conservation Acts) which position conservation as occurring largely separate from any productive land management. While all three of these acts have provisions for commercial use of conservation land, such commercial activities must be secondary to the primary conservation focus of land management (and timber harvesting is specifically excluded). This is of course an important safeguard as much of the land managed under the Reserves, National Parks and Conservation Acts contains our most important and valuable indigenous ecosystems. However, the approach of these acts is philosophically very different from that of the Resource Management and Forests Acts. These latter acts, by promoting integrated land management for both production and biodiversity conservation, provide a more appropriate approach to land management on the 70 per cent of New Zealand that is not public conservation land.

The Timberlands debate: sustainable forest management versus politics

Timberlands West Coast Ltd, a state-owned enterprise, was established in 1990 to manage the state-owned indigenous forests that had been set aside in the 1980s for timber production on the west coast of New Zealand's South Island. While most indigenous forests administered by the NZFS were transferred to the DOC in 1987, the west coast indigenous forests were subject to a major consultative exercise on their future status which resulted in some forests being retained for production. This occurred because of the economic importance of forestry to the west coast region and because the plantation resource was not yet mature enough to sustain the local sawmilling industry without indigenous timber supplies. The consultative exercise, chaired by the secretary for the environment, involved bringing all interested parties (government agencies, local government, environmental NGOs and industry) together to identify which forests should become part of the public conservation estate and which should be retained as production forests. The resultant West Coast Accord, which was signed by all the parties, was seen at the time as a major achievement in resolving environmental disputes. The West Coast Accord provided for the immediate transfer of substantial areas of indigenous forest to the public conservation estate (which now accounted for 78 per cent of the west coast land area), the establishment of a transitional period of unsustainable over-cutting of specified indigenous forests to provide the forestry industry with sufficient timber until the plantation forest resource matured, and the sustainable management of the remaining areas of indigenous forest.

The indigenous forests in this third category included 9500 hectares of lowland podocarp (mainly rimu) forest for which sustainable management was implemented in 1992, and a further 45,000 hectares of beech forest that were going to be managed using a similar approach. The overall goal of Timberlands management for the lowland podocarp forests was to undertake management intervention that is in alignment with spatial and temporal scales of natural disturbance process in the forest ecosystem consistent with information of the day. The key management objectives for meeting this goal were to maintain the pre-harvested state of the forest in terms of biomass, tree size-class range (especially presence of old trees), natural spatial patterns, relative proportions of the major tree species, and forest timber quality.

The sustainable forest management plan written for lowland podocarp forests attempted to balance timber production with ecological sustainability and the public perception of the forest's natural values.[3] In silvicultural terms, the plan aimed to:

- maintain within the managed forest biomass, structure and composition of tree species equivalent to that in comparable undisturbed forest

- harvest a proportion of trees across a range of diameter size-classes according to their predicted mortality rates rather than simply taking trees from a particular size-class (high-grading)

- select trees for harvest that are expected to die within a particular management time-frame

- individually select single trees or small groups of trees (up to 4 trees maximum) and use GPS to ensure an even coverage of the area

The overall aim of management was to maintain a healthy forest ecosystem in terms of stand structure and species composition that includes dead and dying trees, and to continue to sustain the full range of biodiversity that occurs within the forest.

Forest management used a forest growth model that was based on tree numbers rather than volumes, allowing a precise account of the yield to be made during the tree selection and harvesting operations. This made it easier to maintain forest composition and structure, and required for every stump and log to be tagged and a GPS record taken to provide an auditable record of the size and location of all trees felled within the forest. Permissible harvests are determined using a transition matrix model which involved modelling each diameter class separately. The model was based on data on tree size ranges, recruitment rates, growth rates, and mortality rates. One problem with the forest growth model is that it is essentially an equilibrium model in that it aims to maintain the current biomass and structure of the forest. Clearly these forests are non-equilibrium systems that are structured by a mixture of frequent and infrequent disturbance events. However, it was intended to avoid the potential of the model to force the forest into a static condition by recalculating it at regular intervals (eg every

3 James and Norton 2001

ten years) so that it is sensitive to disturbance-induced changes in forest structure.

The permissible yield for the forest was determined by assigning a proportion of the expected mortality over a 15-year felling cycle to harvest. The permissible harvest, expressed in terms of numbers of trees within 10 cm diameter size-classes per year over the felling cycle, was determined individually for five forest-landform types based on resource data specific to that type. Subdivision of the forest by landform was important because of substantial differences in forest composition and tree growth rates on different landforms. For example, average diameter growth rates for rimu 70–80 cm in diameter on better-drained moraines are more than twice those on poorly drained terraces. Finally, natural windfalls harvested, or trees harvested during road construction, were included within the yield so as not to over-cut the forest.

One of the biggest problems with earlier forestry operations in these forests resulted from the high impact of ground-based harvesting operations on forest soils and soil water tables. A major change in harvesting technology proved financially viable when ex-military Russian heavy-lift helicopters became available after the fall of communism in Eastern Europe. The most commonly used helicopter, the Mil 8, has a five tonne lift capacity and when fitted with a log grapple can harvest upwards of 100 tonnes per hour. Helicopters enabled the economic harvesting of a widely scattered log resource with negligible impact on the forest ecosystem, especially soils, and substantially decreases roading needs. The high productivity of the helicopter also means that they are only present in the forest for a few days each year so that disturbance to avifauna is minimised.

While any forest management will have some impacts on indigenous biodiversity, the consequences of the sustainable management approach used by Timberlands are very different from earlier management of these forests by the NZFS. Forest harvesting by Timberlands was based on single tree extraction and the harvesting of only a proportion of trees across a range of size classes, ensuring that there would always be large (old-growth) trees and standing dead trees in the forest to provide habitat for indigenous biota. This is especially important for those birds that require cavities for nesting.

Biodiversity within sustainably managed indigenous forests may in fact benefit from the type of management that was undertaken by Timberlands. Most indigenous forest animals in New Zealand forests are being severely predated by introduced rats, stoats and brushtail possums. Timberlands were controlling, or were preparing to control, these animals as part of their mitigation responsibilities under the RMA. Because the abundance of many indigenous animals, especially birds, appears to be more dependent on predation than on resource availability, animal numbers may in fact be higher in sustainably managed forests with pest control than in a comparable unmanaged indigenous forest without pest control.

The current and proposed sustainable management of west coast indigenous forests by Timberlands became an election pledge during the 1999 central government election campaign, with the newly elected Labour–Alliance coalition government quickly moving to change Timberlands' 'Deed of Appointment' in order to remove beech forest management from their management objectives. Subsequently, government passed legislation (Forests (West Coast Accord) Act 2000) that cancelled the West Coast Accord and exempted the government from paying any compensation for any loss or damage arising from this cancellation. At the same time, this legislation allowed the government to change the tenure of land managed by Timberlands to conservation, reserve, or national park land. This legislation also put in place a timeframe for the cessation of the sustainable management of the west coast podocarp forests. These political changes occurred at the time Timberlands was having its application for resource consents for forest management under the RMA heard by the local authority. The consent hearings would have enabled full public debate over the merits of the proposed management. All the indigenous forests that were being managed by Timberlands—or were proposed for management—have now been transferred to the public conservation estate and are managed by the DOC, including some areas as national park, despite having a long harvesting history.

The criticisms of Timberlands' proposed indigenous forest management by many environmental groups, politicians, and others is a good example of attitudes towards environmental management that have not embraced the land management paradigm found in the Resource Management and Forests Acts. The land that Timberlands was managing or was proposing to manage was not part of the public conservation estate but had been set aside by central government for sustainable timber production through the West Coast Accord. Timberlands' indigenous forest management represented an example of integrated land management where both production and conservation goals could be meet in the same landscape. The sustainable management that was being undertaken by Timberlands in its podocarp forests meets the requirements of the Forests Act for a Sustainable Management Plan (see below), while the results of an initial assessment (undertaken prior to the 1999 and 2000 Government decisions) of this management in terms of the international Forest Stewardship Council (FSC) certification system suggested that Timberlands' sustainable management approach would gain such international environmental certification.

Indigenous forest management today

While the majority of indigenous forests are now part of the public conservation estate, there are still about one million hectares in private ownership, and not subjected to protective covenants, a proportion of which is suitable for sustainable timber production (estimated as up to 500,000 hectares. These forests, with the exception of some indigenous forests under Maori ownership, must be managed in accordance with Part IIIA of the Forests Act which is administered by the Ministry of Agriculture and Forestry (MAF). Part IIIA of this act provides the framework for the sustainable management of indigenous forests and includes provisions to regulate sawmilling and the exporting of indigenous timber and timber products. Part IIIA specifies that any indigenous timber from privately owned indigenous forests can only be milled if it comes from a forest that has a sustainable management plan or permit, or under specified 'other provisions' within the act. Sawmills that are found to have indigenous timber from other sources can be prosecuted. The act also outlines the type of harvesting that can be undertaken (Second Schedule) limiting harvesting to single trees or small groups for podocarps, kauri and 'shade-tolerant and exposure-sensitive broadleaved hardwood species', and to coupes no more than 0.5 hectares for beech and 'other light-demanding hardwood species', and requiring forest owners to replant where natural regeneration is inadequate.

Forest management plans and permits differ in their term and the volume of timber that can be harvested. Reflecting this, they also differ in the amount of information that is required to gain a plan or a permit. Plans that have a 50-year term are registered against the land title of the property and require detailed information on land tenure, forest type and sustainable harvest levels, proposed management including harvesting system, protection measures, and other values. MAF are required to consult with DOC on all plans and with Te Puni Kokiri (Ministry of Maori Development) for applications covering Maori land. Once approved, annual logging plans must be prepared and approved before harvesting can start, while the main plan is subject to five-year reviews. Landowners are required to ensure that any resource consents required by regional and/or district councils are also obtained before commencing operations.

In contrast, sustainable management permits provide for the harvesting of limited amounts of timber within a 10-year period comprising up to 250 cubic metres of podocarp, kauri or shade-tolerant hardwood species, or 500 cubic metres of beech or other light-demanding hardwood species. Permits require less information than plans, although annual logging plans are still required.

To assist with the better implementation of sustainable forest management, MAF has recently published detailed standards and guidelines for forest management with respect to Plans and Permits. These have been developed through an extensive consultation exercise to ensure that they will meet the requirements of the Forests Act. The standards and guidelines are organised hierarchically and include criteria, goals, indicators, benchmarks, and verifiers. In addition MAF has commissioned research on the development of a forest-site classification system to help forest owners in better planning their forest management to take into account spatial heterogeneity within their forests.

At present approximately 80,000 hectares of private indigenous forest are covered by approved sustainable management plans and permits and as such this can be seen as a 'boutique' industry compared to the dominance of plantation forestry within New

Zealand. Nonetheless, the remaining indigenous production forests are now providing a sustainable source of specialised timber that is much sort-after for uses as diverse as furniture and axe handles. Furthermore, two indigenous forest owners have achieved FSC certification for their forest management.

Plantation forest management

The goal for the management of New Zealand's rapidly expanding plantation forest estate has historically been seen as primarily one of producing a sustainable supply of wood to replace the declining indigenous wood supply (see figure 10). While the goal was one of sustainability, the focus was primarily on the sustainability of wood supply rather than on sustaining broader environmental values. However, there was a widespread view that the simple presence of plantation forests did meet broader environmental goals because they provided an alternative to clearing indigenous forests for timber. With the demise of the NZFS in the late 1980s and the end of most indigenous forestry, some environmental NGOs began to scrutinise plantation forestry. This resulted in 1991 in the signing of the New Zealand Forest Accord by both industry and environmental NGOs. In signing the New Zealand Forest Accord (and subsequent Principles in 1995), the forest industry recognised that the existing indigenous forests should be maintained and enhanced and that areas of naturally occurring indigenous vegetation would not be cleared for plantation establishment. In return, the environmental NGOs recognised the important contribution that plantation forests make to the New Zealand economy and the role that these forests played as an alternative to the depletion of indigenous forests.

While recognising that both indigenous biodiversity conservation and plantation forestry have important roles to play in New Zealand, the New Zealand Forest Accord seeks to effectively separate production from biodiversity conservation, viewing plantation forests as crops that do not need to meet biodiversity conservation goals in themselves. While there is general agreement about the relative importance of plantation forestry and indigenous forest conservation in terms of the New Zealand Forest Accord, there is a growing recognition that plantation forests can and should contribute more to biodiversity conservation than just providing an

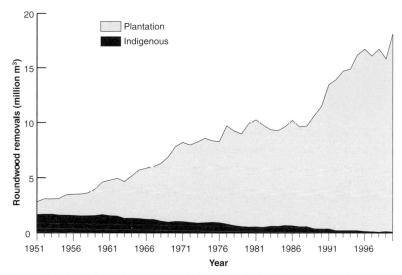

Figure 10 Estimated roundwood removals from New Zealand indigenous and plantation forests (Ministry of Agriculture and Forestry statistics)

alternative timber supply to indigenous forests. This case was argued strongly by Rosoman,[4] although many at the time saw it as representing a major handicap for plantation forestry. While acknowledging that trees and forests are essential to nature and human society, Rosoman argued that plantation forestry had a number of significant impacts on soil and water quality, yield, indigenous biodiversity, and ecosystem health that the industry needed to address.

While some in the industry still regard plantations as essentially crops, there is increasing acceptance among plantation forest managers that broader environmental issues including biodiversity need to be addressed in plantation management. A position statement by the New Zealand Institute of Forestry highlights this shift in attitude. This statement emphasises the importance of considering biodiversity 'in both planted and natural landscapes' and 'the importance of identifying safe and efficient operational practices to enhance and maintain biodiversity in a productive environment'.

---

4 Rosoman 1994.

Plantation forests usually have less habitat diversity and complexity than indigenous forests, and plantation management practices such as clearcutting, and pesticide and fertiliser use all adversely affect indigenous biodiversity in and adjacent to the plantation. Nonetheless, plantation forests do provide suitable habitat for a range of indigenous species, and compare favourably with most other land uses involving primary production. For example, plantation forests provide habitat for a wide range of indigenous plant, animal, and fungal species including, in northern New Zealand, habitat for the North Island brown kiwi. However, the abundance and composition of indigenous species is dependent on stand age, as well as topography, aspect, soil nutrient and moisture status, silvicultural history, land-use prior to plantation establishment, and proximity to indigenous forest remnants.

While plantation forests provide important habitat for many indigenous species, they are also important for biodiversity conservation at the landscape level. Plantation forests can be important as buffers to indigenous forest remnants. For example, indigenous forest-edge microclimates have been shown to be much more similar to forest interior microclimates when the adjacent land use is plantation than when it is pasture. Furthermore, the vegetation composition of indigenous forest remnant edges is more similar to the remnant interior when the edge is adjacent to plantation forest than when it is adjacent to pasture. The presence in plantations of a number of indigenous birds normally restricted to indigenous forests—especially where the plantation occurs close to indigenous forest remnants—suggests that plantations can also improve connectivity between indigenous forests for these species.

Reflecting these changing views about the role of plantation forestry in sustaining a diversity of environmental values, including indigenous biodiversity, New Zealand forestry companies attempted to develop their own system for environmental certification of plantations. However, international market pressure caused plantation managers to adopt the FSC certification system. Nine plantation forestry companies have now (mid 2002) received FSC certification comprising some 30 per cent of the total plantation area, and further companies, including the largest New Zealand forestry company, are currently being assessed or working towards FSC assessment. While a greater awareness of environmental issues is an important incentive for certification, a number of other factors are also important including enhanced market access and financial returns, contribution to triple bottom-line reporting, and facilitating future resource consent applications (eg for new plantings). While it would be naive to think that plantation forestry companies are seeking certification for environmental reasons alone, the greater awareness and consideration of environmental issues in plantation management is making a positive contribution towards indigenous biodiversity conservation in these working landscapes.

While certification is an important step towards sustaining indigenous biodiversity, it is only the beginning rather than the end of the process. Many plantation companies meet their certification requirements by implementing management of the indigenous remnants within their plantations (eg through animal pest control) and by ensuring that the environmental impacts of their plantation management (eg harvesting) are minimised. However, if plantation forestry is to be truly sustainable with respect to indigenous biodiversity, then it needs to look more broadly at how forestry operations interact with biodiversity at the scale of the landscape.

There are several ways in which management can be adapted to help sustain indigenous biodiversity within plantations.[5] For example, several studies have suggested that the establishment of a greater diversity of planted species will increase the range of habitat types available for indigenous species, and that an increase in rotation length will enhance indigenous biodiversity. Adoption of alternative silvicultural systems such as continuous cover forestry and variable retention harvesting also have considerable potential and warrant further investigation. Important to any new approach to plantation forest management is the need to take a landscape approach to such management. Plantations can be considered to comprise a spatial array of different elements that can be arranged in different ways depending on management goals. The key elements within a plantation forest are individual stands or compartments of different age and species composition, remnants of indigenous ecosystems, including riparian strips, and amenity plantings. Some of these elements are fixed in the landscape (eg indigenous remnants

5 Norton 1998; Hartley 2002.

and riparian strips) but others can be arranged in different ways. In North America, spatial modelling tools have been used to optimise timber harvesting in indigenous forests to meet biodiversity conservation goals. Similar modelling could be used to optimise the arrangement of different aged plantation forest compartments, and different plantation species, with respect to indigenous remnants and other adjacent land uses to maximise both timber production and biodiversity conservation.[6] It would seem likely that within a few decades plantation managers may need to adopt many of the principles being advanced in indigenous forest management.

Reflecting on the New Zealand experience

A number of conclusions can be drawn from New Zealand experiences with sustainable forest management, both with respect to indigenous and plantation forests. None of these are necessarily unique to New Zealand, but the politically driven changes in New Zealand land management have certainly thrown them into sharp relief.

As has been the case in other countries, the long-term survival of indigenous forest management on publicly owned lands is totally dependent on political support. Even though Timberlands was implementing sustainable forest management, political considerations overrode this in the decision to end government's direct involvement in timber production from indigenous forests. Although there were attempts in government at the time to claim that this decision was based on scientific considerations, the key political players eventually admitted that it was a simple political decision based on philosophical rather than scientific concerns.

The Timberlands decision was, however, not only about what was happening in the late 1990s. It was the final chapter in a series of public forest conservation debates that stretched back to the 1950s. It would seem reasonable to speculate that no matter how sustainable Timberlands operations were, there was still sufficient mistrust of government forestry (especially from NZFS days) and too much memory of past conflicts for environmental NGOs to

---

6 Norton 1998.

have left Timberlands alone. Many of the players involved in the often bitter conflicts between the NZFS and environmental NGOs in the 1970s and 1980s also played a decisive role in the 1999 and 2000 government decisions.

It is, however, important to recognise that New Zealand was able to move away from indigenous forestry because of the dominance of plantation forestry and through the importing of high-value timbers, often from forests elsewhere in the world, not all of which are sustainably managed. Without the plantation forest resource, the nature of New Zealand forestry would be very different today to what it is now.

The demise of government involvement in indigenous forestry has resulted in a major opportunity for private indigenous forest owners as timber prices have risen rapidly as state subsidised harvesting has ended. For example the average price of rimu at the mill rose from $NZ150 in 1993 to $NZ375 (and as high as $NZ500) in 2001.[7] However, environmental NGOs are suspicious of sustainable forestry on private land and forest owners will have to ensure that they are able to demonstrate that they can achieve international benchmarks, such as FSC certification, in their management to ensure the survival of this industry.

The attitude of the plantation forest industry to environmental issues has changed rapidly over the last decade, primarily in response to the perceived market advantages of gaining FSC certification. However, industry needs to recognise that certification is only the beginning of the process, not the endpoint, and that society will continue to watch forest management closely and expect forest managers to respond to their concerns. Many in the plantation forest industry have questioned why they should come under such scrutiny compared with, for example, the agricultural sector. The question is valid, but it doesn't change the reality of the public perceptions that forest managers face.

The exciting challenge for plantation forest management is to expand the focus of sustainable management away from the stand or even forest and to consider how forestry can contribute to indigenous biodiversity at the landscape scale. In a previously

---

7 Griffiths 2002.

forested country like New Zealand, plantation forests have much to offer indigenous biodiversity when the alternative land use is usually pastoral. Plantation forests, if properly managed, and especially with consideration to spatial landscape issues, can contribute significantly to biodiversity conservation in an environment in which indigenous remnants are usually a minor component.

Acknowledgments

My thanks to Rob Allen, Alan Griffiths, Ian James, Bruce Manley, Euan Mason and Craig Miller for constructive comments on a draft of this manuscript.

11

A new forest and wood industry policy framework for Australia

Judy Clark

Australia can now meet virtually all its sawn timber and paper needs from already established plantations (agricultural tree crops) and recycling, and it can also choose to switch all its native forest woodchip exporting, now the main market for native forest wood, to hardwood plantations. Plantations offer a pragmatic solution to Australia's forest problem because they can substitute for virtually all native forest wood uses. However, despite Australia's rapidly increasing plantation supply in the 1990s, there has been no overall reduction in native forest logging. This suggests that Australia's forest problem is not an unsolvable clash of environment versus industry development objectives, but rather a lack of political will to solve old conflicts.

Unfortunately, most of Australia's increasing plantation supply in the 1990s was exported unprocessed as whole logs or woodchips, which means that opportunities to build rural economies and manufacturing employment—through processing plantation wood domestically—are being lost. Australia's forest problem is now a combination of lost biodiversity and native forest habitats and lost plantation-based employment and manufacturing in rural areas.

The 'multiple use' approach to native forest management for wood production, which retains its stronghold in Australia, is fundamentally flawed because it ignores the reality of the price-cost squeeze of commodity production. There is an alternative approach to framing forest policy that enables Australia to solve its forest conflict while, at the same time acknowledging and addressing the price–cost squeeze of commodity production. Strategies to enhance the persistence capacity of native forests as self-sustaining ecosystems, wood production systems to meet our material needs and rural socio-economies are not in conflict, but rather, they work positively across the core interests. The key to coherent policy is bringing the plantation resource directly into the framework.

An evaluation of the performance of the Australian wood and wood products industry and government policy during the 1990s using these strategies as the criteria finds that Australia is dismissing major opportunities for native forest conservation, investment in plantation processing, and rural manufacturing employment.

Cost reduction

Cost reduction is the main game for producers of most wood and wood products. Borshmann, in his oral history of people's interaction with the Australian bush, captures its reality for Andy Padgett, a Tasmanian logging contractor.[1] Padgett claims to be the first person to fell a tree in Tasmania with a chainsaw. Rival cutters quickly copied him, and soon axes and cross-cut saws became redundant. Now chainsaws are being replaced by feller-bunchers— a one-person operated machine that fells, debarks, and loads logs. Padgett brought the first Kenworth truck to Tasmania in 1965, and its braking system reduced log haulage time to a third. Soon other truckies had the same brakes. Padgett, with others, developed the folding skeletal log trailer that on return trips to the forest could be folded in the middle and loaded back onto the truck, saving on fuel and tyres. The trailers are now sold worldwide. Padgett's cost reduction strategies are motivated by profits, the lifeblood of competitive market economies.

---

1 Borschmann 1999.

Such logging industry cost reductions have environmental consequences on both the supply side and the demand side of the market system. On the supply side, previously inaccessible and distant native forests became commercially viable to log. The combination of modern logging equipment and new markets for eucalypt chiplogs in the Japanese pulp and paper industry brought about the introduction of clearfelling to Australia's native forests. Clearfelling enabled logging contractors to spread the cost of their expensive machinery over the larger revenue from the increased cut, thereby reducing logging costs per unit of wood. Although new technologies can be softer on the environment, for example the replacement of crawler tractors with rubber-tyred machines, their adoption is largely motivated by the embodied cost saving.

Industry employment is also a casualty of cost reduction. Today, Australia's logging industry (felling and carting to mill) employs two-thirds the number of people employed before World War II, even though it logs four times the volume of wood. Labour productivity (production per person employed) has increased by an average 3 per cent per annum over the past six decades.

Cost reduction also occupies the minds of those selling native forest wood. In Australia they are largely state government bureaucracies managing forested public land. Whether or not forest agencies are puppets of industry, they are now more directly exposed to the cost-reduction pressures of commodity production because they are increasingly managed on commercial business lines. Activities aimed at environmental enhancement (eg pre-logging flora and fauna surveys, increased rotation lengths, and additional buffer protection) add cost and usually reduce wood supply. They fundamentally conflict with the forest agency's commercial objectives.

On the demand side are the consumers—the buyers of intermediate products (eg wood, woodchips, and pulp) and final wood products (eg sawn timber, paper, and wood panels). Buyers of intermediate products seek out or try to negotiate the lowest price as part of their cost reduction strategy. The pressure passes down the line to the logging contractor, the forest agency and, ultimately, the native forest environment itself. The final consumers are the ordinary householders. We characteristically buy the cheapest of the highly standardised products on offer. Companies making the products compete for our dollar and the producer of the cheapest product usually wins. Cost-cutting passes down the line, ending at the environment.

This combination of cost-conscious producers and price-conscious final consumers embodies the fundamental flaw in the 'multiple use' approach to native forest management for wood production. Rarely is the nature of commodity production in competitive market economies and its implications for native forest ecosystems acknowledged in policy.

Wood as a commodity

Over the past five centuries, the word 'commodity'—from the Latin *commoditas*, meaning benefit—has remained loosely defined. Originally 'commodity' referred to imports from 'newly discovered' lands (eg spices, cottons, silks, furs) and to the traditional commercial products of farms. Marx gave 'commodity' a new meaning, namely, products made in factories to sell at a profit in the market. Financial writers today use 'commodity' to refer to market-traded products like oil, coal, gold, wood, and sugar that undergo varying degrees of processing. These loose understandings call for a definition that draws out the distinguishing nature of a commodity.

Both commodity products and specialty products, their opposite, are sold in the market. Commodities are different from specialties in that they are homogeneous products that usually meet established standards. Most airline business is commodity based, as is much tourism. Commodities dominate our supermarket purchases. Homogeneity in commodities leaves little, apart from price, on which commodity producers compete for the buyers' dollar.

Price-dominated competition characterises commodity production. To capture more sales, commodity producers focus on selling their products at attractive prices, thereby putting downward pressure on commodity prices over the long term. Real commodity prices today are a mere fraction of what they were in the industrial age and the agrarian age before that.[2]

Declining commodity prices are a widely understood reality in the wood and wood products industry, although some analysts phrase their findings more cautiously—there is no evidence of

2 Ruthven 1995 provides a lucid account of the trend.

increasing real prices for wood—and others project mildly increasing real prices for globally traded wood.[3] To maintain profit levels in an environment, or expected environment, of real prices in long-term decline, commodity producers adopt an ongoing strategy of cost reduction. Technological change (new ways of making products and new products) is paramount.

The wood and wood products industry is highly exposed to the price–cost squeeze because of its highly commodified nature. Globally, commodities make up around 80 per cent of the output volume of the wood products industry and an estimated 95 per cent of Australia's production. Most sawn timber, structural beams, and wood-based panels, much of veneer and plywood, pulp and most paper are commodities. The wood used in their manufacture is highly commodified. Most sawlogs and all chiplogs and woodchips are commodities.

Implications for native forest ecosystems and multiple use

Old-growth native forests provide the wood products industry with a huge stock of raw material. The cost-reduction effort initially concentrates on lobbying for lower stumpages (price paid for logs before felling) and developing technologies to reduce logging and transport costs. As the old-growth native forest resource diminishes, the cost-reduction effort shifts to the economics of growing wood in an agricultural regime. It offers cost reductions through economising on time, land, logging, and transport as well as scale economies and processing efficiencies.[4] The potential for ongoing cost reduction, through research and development in tree breeding and plantation management, sustains industry interest in agricultural wood growing regimes.

The price–cost squeeze of commodity production means that users of native forest wood can be expected to lobby for intensification of native forest management in order to enjoy the same commercial advantages as their competitors using an agricultural

---

3 See Leslie 2001; Clark 2001; Humphreys 1992 for the declining real price expectation. See Food and Agricultural Organisation of the United Nations 1997; Sedjo & Lyon 1990 for a more cautious presentation. See Sohngen et al 1999 for the increasing real price expectation.

4 For a fuller account, see Sedjo 1983; Clark 1995.

grown crop. Opportunities for applying intensification technologies are greatest when old-growth forests are clearfelled and shift into a regrowth phase. Intensification practices, which include reducing rotation lengths, increasing wood yields using agricultural technology, and selecting non-indigenous/endemic tree species for replanting, threaten biodiversity in native forest ecosystems.

'Multiple use' has been the preferred approach to native forest management for wood production in Australia. It is a loosely defined concept that has always sat uneasily with the main role of public foresters to supply wood. From a forestry viewpoint, it was vital that the 'water-sensitive' Australian public accept 'multiple use' of forested water catchments because Australia's more productive native forests were located in water catchments. By the 1950s, foresters globally were becoming increasingly concerned about public pressure for the alienation of native forests from wood production. The fifth World Forestry Congress held in Seattle in 1960 adopted the theme 'Multiple Use of Forested Land' and Richard McArdle, Chief of the USA Forest Service, in his keynote address spoke of forested land being excised for non-wood purposes. Of great concern was the increasing pressure to set aside additional forested land exclusively for recreational use. McArdle emphasised the need to apply multiple-use management widely and intensively 'to lessen the pressures to divert forest lands'. It was hoped that a multiple-use approach could reduce or resolve conflict by establishing a balance in the competing uses for native forests. The 'multiple use' of native forests appears, however, to be unable to meet the needs of commodity producers facing the price–cost squeeze of commodity production without threatening the ecological integrity of native forests. Trade-off is required and conflict is inevitable.

Forest and wood industry policy framework

Australia's forest and wood industry policy was subjected to political and bureaucratic scrutiny during the 1990s, but this work largely ignored the implications for industry, employment, and the environment of the price–cost squeeze of commodity production. Here I propose an alternative forest and wood industry policy framework that explicitly acknowledges the reality of commodity

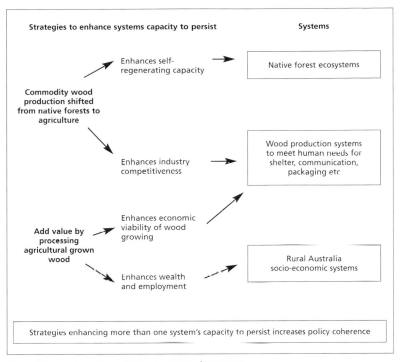

Figure 11 Forest and wood industry policy framework

production.[5]

The first task is to identify the systems society would like to survive or persist for a designated time or life span. For our interest—forests and wood production—I select three systems:

- native forests as self-regenerating ecosystems
- wood production systems to meet human needs for shelter, communication, packaging etc
- rural socio-economic systems

This selection is based on my assessment of the Australian public's desire to protect native forests; meet its requirements for shelter, packaging, and communication; and also maintain rural economies and employment. The analysis can be readily extended to include

5 I build on the approach to sustainability developed by Costanza & Patten 1995.

other systems, the most obvious being the atmosphere.

The second task is to develop strategies to enhance the persistence capacity of the systems, without major trade-off. This essay proposes two key strategies:

- shifting commodity wood production from native forests to an agricultural system
- adding value by domestic processing

Shifting commodity wood production from native forests to agriculture
Shifting wood production from a self-regenerating native forest ecosystem to plantations—an agricultural system—has the combined effect of enhancing the persistence capacity of native forest ecosystems as well as the economic viability of the wood and wood products industry. This proposal is sometimes dismissed because of an incorrect understanding that more plantations are required for such a substitution. The substitution option, however, does not require one more tree to be planted in Australia. Our existing softwood and hardwood plantation estate covering 1.5 million hectares can now meet virtually all Australia's wood requirements and, within a couple of years, can completely substitute for native forest hardwood chip exports.[6] Australia's existing plantations are expected to supply around 700 million cubic metres of wood over the next 30 years. The question is, does Australia want to use this existing resource to enhance industry efficiency and also secure immediate and significant native forest conservation?

We should not ignore the choices presented by existing plantations because of environmental and social concerns enveloping extensions to the plantation estate. They are largely separate decision-making exercises. Fortunately, interest is building in moulding future tree planting to address land-degradation problems flowing from earlier extensive native forest clearing for agriculture and also to develop environmental safeguards for biodiversity and water.

---

6 See Clark 1995 for Australia's plantation substitution potential and Ferguson et al. 2002 for the latest government published projections of plantation wood availability.

Adding value by domestic processing

'Value added' measures the economic value generated by a firm or industry in using land, labour, capital, and knowledge to make products. It is the difference between the total value of a firm's production (output times price) and the cost of all the purchased material inputs and services that it uses.

Income (as measured by value added) per unit of raw material usually increases with further processing. I calculate that processing sawlogs domestically into sawn timber, rather than exporting them unprocessed, boosts income per cubic metre of wood used sevenfold. Income per unit of wood used is even greater if the processing of sawmill residues into wood-based panels is included in the calculation. The income returns per cubic metre of wood used by the Australian paper industry are even greater (see figure 12). This is because the Australian paper industry's high use of recycled paper has significantly increased its resource productivity (output per unit of raw material input). Significantly higher levels of income per cubic metre of wood input are realised when commodity sawn timber, panels, and paper are further processed into wooden components, roof trusses, door frames, paper containers, and so forth.

An industry that undertakes a high level of domestic processing (ie wood growing and processing is highly integrated domestically) generates more regional/national income per unit of wood grown than an industry that exports most of its wood unprocessed as logs or chips. This means that, from the perspective of Australian industry policy (leaving the issue of comparative advantage temporarily to one side), a highly integrated domestic industry has more scope and more items to factor into its cost reducing strategies relative to an industry that exports most of its wood unprocessed. Moving along the further processing path—shifting into semi-commodity and specialty products—also reduces the intensity of price competition and therefore the cost-reduction imperative.

Policies aimed at facilitating greater domestic integration in the wood-based industries require attention to the cost and revenue sharing arrangements between plantation growers and processors. The issue creates tensions in the wider agricultural sector and caused intense debate among Australia's early foresters. A calmer debate in today's environment of abundant plantation wood availability could make a valuable contribution to wood and wood products industry policy.

Industry structure has significant socio-economic implications for rural communities in close proximity to plantations. I calculate that processing wood into sawn timber, panels, and paper generates

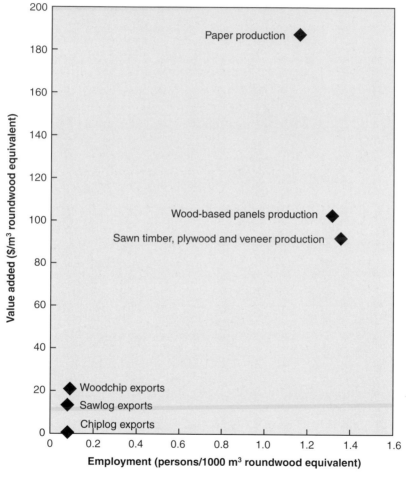

Figure 12 Value added and employment per unit of wood used in the Australian wood and wood products industry. (Value added is revenue less costs. Employment is that generated inside the mill gate, and excludes employment that is general to all sectors of the industry, namely wood growing, logging, and haulage. Roundwood equivalent is the volume of wood in log form required to produce the wood product. *Source:* Clark 2002.)

around fifteen times more jobs than exporting the equivalent log volume unprocessed (figure 12). Industry policy that supports domestic processing of agricultural raw materials should be a high priority for those seeking higher employment levels and increased economic wealth in rural Australia.

A wood and wood products industry policy aimed at securing a highly integrated domestic industry structure is anathema to those advocating free-market policies based on comparative advantage. It is commonly perceived that Australia's comparative advantage lies more in agriculture (in this case, growing wood) than in manufacturing (processing wood into wood products). The evidence suggests, however, that Australia's comparative advantage lies more in manufacturing certain wood products than in wood growing. An international plantation benchmarking study undertaken in 1995 by forestry consultants Margules Groome Pöyry found that Australia's plantation wood-growing cost competitiveness is lower than all competing countries except New Zealand. In wood processing, however, the Industry Commission in 1993 reported that Australia was globally competitive in packaging and industrial papers, potentially globally competitive in hardwood kraft pulp, and had a slight disadvantage in particleboard, medium-density fibreboard, softwood plywood, and softwood sawn timber on export markets. These findings encourage Australia to identify and address the barriers to investment in plantation processing, rather than complacently sticking to the view that Australia's comparative advantage stops at the farm gate.

Evaluating Australia's performance

The two strategies to enhance the persistence capacity of the three identified systems can be readily formed into indicators of improvement or deterioration in contending with the price–cost squeeze of commodity production. The two indicators are:

- the shift in commodity wood production from native forests to plantations

- domestic integration in plantation wood growing and processing

Neither the Australian Bureau of Statistics nor the Australian Bureau of Agricultural and Resource Economics provides data that enables the indicators to be measured directly. This is because wood and wood products economic activity is not comprehensively reported by growing regime (plantation or native forest based). Nevertheless, government published data can be readily disaggregated by growing regime using published and unpublished sources.[7] This exercise was undertaken for 1999/00 and, to enable a comparison over time, 1989/00.

Shift in commodity wood production from native forests to plantations
With a near doubling in plantation wood supply, Australia's wood products industry (producing sawn timber, wood panels, and paper) substantially increased its plantation dependency over the decade ending 2000. Approximately 75 per cent of Australia's production of wood products was made using plantation wood in 1999/00— only 25 per cent was made using native forests.

Despite significant displacement of native forest wood in domestic processing, there has been no decline in the native forest cut. Native forest logging remained constant at approximately 11 million cubic metres per annum over the decade ending June 2000. Production was maintained by increased exports of native forest woodchips. That native forest logging remains unchanged even though plantation supplies have increased significantly indicates that governments are dismissing opportunities to reduce the commodity wood production threat to native forest ecological integrity.

Integration in plantation wood growing and domestic processing
The Australian plantation sector has become less domestically integrated over the decade ending 2000. At the start of the 1990s, virtually all Australia's plantation resource was processed domestically. At the end of the decade, one-third was exported unprocessed as chips or whole logs. Exports of unprocessed wood accounted for nearly two-thirds of Australia's increased plantation wood production over the 1990s. Domestic processing increased, but its growth was outstripped by unprocessed wood exports.

---

7 See Clark 2002.

The softwood sector, which overwhelmingly dominated plantation wood supply during the 1990s, accounts for most of the deteriorating domestic integration. Softwood chip exports increased significantly during the first half of the 1990s. The second half of the 1990s saw a leveling out in chip exporting and a rapid increase in softwood plantation log exports.

Policy implications

Australia is not enjoying anywhere near the full economic, social, and environment benefits of its plantations. Most of the doubling in Australia's plantation supply over the 1990s was exported unprocessed as whole logs or woodchips, taking with it significant potential investment in rural economies and manufacturing employment. There has been no reduction in native forest logging over the 1990s, despite Australia's increasing plantation supply that can substitute for most native forest products.

The alternative approach to forest and wood industry policy for Australia presented here specifically acknowledges the price–cost pressures of commodity production and specifically includes an agricultural wood-growing regime that jointly accommodates the interests of commodity producers and native forest ecological integrity. Within this framework, key strategies to enhance the persistence capacity of native forest ecosystems, wood production systems and rural socio economic systems were found to be not in conflict. The framework enables a high degree of policy coherence because each strategy enhances the persistence capacity of more than one system and, importantly, both strategies enhance wood as a production system.

The task still facing Australia is to align native forest and wood industry objectives to the appropriate land base. This is made easier because commodity production accounts for 95 per cent of the nation's wood-based industry and the existing plantation estate can substitute for most native forest products be they consumed domestically or exported. On the basis of the framework presented here, virtually all wood production should be aligned to plantation land and superimposed by an industry policy encouraging investment in processing. This frees native forests from commodity wood production, thereby removing a threat to its ecological integrity.

Removing barriers to plantation processing investment should become the highest government policy priority for Australia's wood-based industries. Rural Australia reaps relatively secure manufacturing employment from a plantation-based industry with its increasing wood supply and superior competitiveness. A large plantation-processing industry advancing into export markets also enhances the market opportunities for plantation growers. The high substitutability between plantation and native forest commodity wood products links these employment and industry attractions with environmental objectives. Australia can choose to use plantations to meet its commodity wood needs and to expand its wood and wood products export businesses, leaving native forests to meet conservation objectives.

Most Australian governments still persist with policies that undermine the commercial viability of the plantation sector. Native forest wood pricing polices—generating very low and sometimes negative returns to public native forests managed for wood production—handicap the emerging plantation processors because invariably the plantation industry pays more for similarly graded logs. The plantation industry is further damaged by government 'compensation' accompanying decisions to protect native forests or reduce logging rates where it props-up the remaining native forest businesses with grants for capital expenditure and other subsidies.

Fortunately, some Australian states have operated outside the national policy agenda that has largely dismissed the plantation opportunity. The Queensland Government, the environment movement (represented by the Australian Rainforest Conservation Society, Queensland Conservation Council, and the Wilderness Society), and industry (represented by the Queensland Timber Board) agreed in 1999 to initiate a complete transition to plantation processing in southeast Queensland and to protect immediately a large area of public native forests. The Western Australian government decision in 2000 to immediately protect all old-growth forests combined with the decision by the state's sole exporter of native forest woodchips to quickly shift its chiplog supply to plantations, means Western Australia is now undergoing a substantial switch to a plantation-based industry. These actions indicate that there are immediate, pragmatic solutions for Australia's forests if governments decide they want to solve old conflicts.

Pecuniary interests

The author is a shareholder in Australian Ethical Investment Ltd. and has superannuation with Unisuper Management Pty. Ltd., both of which may have investments in companies growing and processing plantation wood.

12

Transitions to ecological sustainability in forests— a synthesis

David Lindenmayer and Jerry Franklin

The modification of forests could be considered one of the defining features of human societies. Some of these human-caused changes to forest resources also have catalysed major events in human history. For example, a major reduction in forest area and an associated shortage of wood is believed to have stimulated the expansion of the Roman Empire.

Changes in forest cover are not always unidirectional, however. Deforestation and other changes in forest cover and condition may be reversed. Indeed, the concept of transitions in forest management is not new. In fact, Mather (1990) proposed a model of changes in forest cover and associated human perceptions based on key historical trends in Great Britain (table 3).

Mather's simple model can be applied to other nations and regions within nations, which may, at a point in time, differ in their stage of forest change. While overall trends in forest area are the primary focus of this model, there are other related characteristics and transitions, such as in how a stable or increasing area of forest estate is being managed.

Table 3 Simple sequential model of forest change

Stage	Description	Trend in forest area	Public perception of Trend
1	'Unlimited resource'	contraction	positive or neutral
2	Depleting resource	contraction	negative
		↓	
		Forest transition	
		↓	
3	Expanding resource	restoration/expansion	neutral/negative
4	Equilibrium (?)	stability (?)	positive/neutral/negative (?)

Source: modified from Mather (1990)

This book focuses primarily on transitions to ecologically sustainable management of forests in developed temperate regions of the world. These changes are being driven primarily by societal demands for management of an essentially stable or declining forest base so as to sustain biological diversity and higher levels of many ecological services as well as the production of commodities. Development of significant additional scientific knowledge about forest ecosystems and biota, including the impacts of and recovery from natural disturbances, have also driven changes in demands as well as contributing to new management alternatives. Transitions to sustainability in management of forest estates entail shifts from a primary emphasis on production of wood products to equal or greater consideration of other values, such as watershed protection and conservation of biodiversity. These transitions are not simply national but are, in fact, increasingly linked to international agreements related to trade, the environment, and product certification. Case studies in this book in which companies had major financial challenges as a result of sustainability issues exemplify these challenges (see Bunnell & Kremsater, chapter 6; Niemela, chapter 7).

Fibre farms and native forests—a land allocation solution to forest conflicts?

One approach to ecological sustainability currently being explored is to move entirely from forest harvesting in native forests to dependence on intensively managed plantations—fibre farms—to produce of timber, pulp and paper, and other wood products. In effect, under this approach, conflicts in forest values are resolved by attempting to isolate the ecological and production functions of the forest on different parts of the forest estate. Forest allocation is often viewed as the simplest and most efficient approach because it reduces the need for compromises among forest values and inter-actions among stakeholders and, consequently, costs. Furthermore, allocation is often viewed as most socially acceptable because of the perceived certainty in land tenure and associated human activities for at least the short term. However, the land allocation approach has numerous problems not the least being that an approach to forest conservation based solely on reserves is insuf-ficient; appropriate management of unreserved areas is also necessary to achieve ecological objectives (Perry, chapter 8).

In New Zealand, the shift from harvesting in native forests to plantations of exotic species is essentially complete, although what was arguably one of world's most ecologically sustainable native forest harvesting operation was eliminated as part of the process (Norton, chapter 10). A greatly reduced area of native forests contributed to the transition—a result of massive earlier clearing of original forest cover throughout New Zealand. In chapter 11, Clark makes the case for a similar transition from native forest harvest in Australia based on current and future availability of plantation resources; she calls for greater investment in Australia's plantation sector, including off-shore processing, and foresees reduced ecological impacts on native forest ecosystems.

Problems and limitations of an allocation-based approach

There are some significant issues that are either directly or indi-rectly associated with a shift in wood production entirely to inten-sively managed plantations, despite the strong impetus of globalisation with its emphasis on production of commodities at the lowest per unit cost (see Franklin, chapter 1: Clark, chapter 11). First, in both Australia and New Zealand, a large quantity of plan-tation material is being exported in an unprocessed state. This dramatically limits employment within the country where the forests are grown and reduces the in-country return on plantation investments (Clark, chapter 11). Further, export of unprocessed wood has the potential to also 'export' pests and pathogens with the wood—thereby creating major environmental problems elsewhere (Franklin, chapter 1).

The need for continued management of native forests—even when they are no longer used for timber production—is a third important complication. Active human management is imperative in many native forest ecosystems in order to achieve important goals for biodiversity conservation and other ecological and social values; simply eliminating forest harvest and declaring that native forests are 'preserves' may be grossly insufficient or even destructive of these values (for several examples see Franklin, chapter 1). Challenges (among many) include:

- accelerated habitat restoration[1]
- reduction of uncharacteristic fuel loadings and restoration of fire to many forest ecosystems in drier regions
- control of pest species, such as feral predators and herbivores in Australia and New Zealand or, conversely, the managed reintroduction of native predators in other jurisdictions, such as North America and Europe (see Soulé, chapter 5)

Shifting forest harvesting from native forests to plantations is currently not feasible in some regions where the plantation base is currently limited, such as British Columbia and parts of Scandinavia. Even in parts of Australia (eg Victoria and Tasmania), major wood products industries are based on harvest of native forests, and transitions to total dependence on plantations are likely to be slow, despite the compelling arguments made by Clark (chapter 11). In such areas, transitions to ecologically sustainable forestry are likely to initially involve changes in management regimes, including silvicultural practices. Changes in management regimes are critical because allocation of the forest estate in these regions into reserves and dedicated production areas will be insufficient to meet conservation objectives for biodiversity and ecological functions (Niemela, chapter 7; Perry, chapter 8). Creating reserve systems that are truly comprehensive, representative, and adequate for all elements of diversity, let alone ecological processes,

---

1 See Carey et al 1999 regarding the need for biodiversity thinning regimes in northwestern North America.

is simply not possible in most temperate regions. Scandinavia is a stark reminder of major limitations of a reserve-only focus for biodiversity conservation with its very limited areas of native forest reserves and major native-forest-based industries (Angelstam, chapter 9; Nielmela, chapter 7).

Transitions to sustainability and the 'variable retention harvest system'

Examples of transitions to ecological sustainability in the regions featured in this book include various management and conservation strategies at the regional, landscape, patch, harvest unit, and individual tree scales. This reflects the multi-scaled challenge of integrating and maintaining multiple economic, ecological and social values of forest. While the general (strategic) objectives are similar, the specific techniques vary among jurisdictions.

Development and testing of silvicultural techniques that are alternatives to traditional clearcutting (clearfelling) does appear to be one of the common themes among all regions, however. The desire is for harvesting and regeneration systems that meet a broader range of objectives than simply wood production. Such altered cutting regimes can be fitted comfortably under the umbrella term of the 'variable retention harvest system'. This system emphasises the structures and organisms that are retained on the harvest unit as much as what is removed (figure 13). An important attribute of the system is that it is not based on replacement of one cutting method by another single prescriptive approach; rather, it provides a conceptual basis for an infinite array of silvicultural prescriptions customised to forest and landscape conditions and management objectives, including worker safety. This flexibility is consistent with the notion that it may not be necessary to maintain **all** elements of the biota on **all** hectares of harvested land (see Perry, chapter 8).

Knowledge of natural disturbance regimes can be used to guide design of silvicultural prescriptions as well as to appropriately place them on the landscape (Norton, chapter 10). For example, in Victorian montane ash forests, higher levels of structural retention to meet biodiversity objectives may be appropriate on areas where natural disturbance regimes (fire, in this case) tended to result in

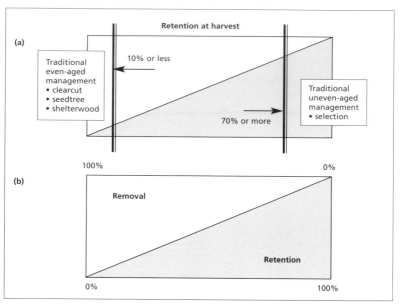

Figure 13 The variable retention harvest system (redrawn from Franklin et al 1997)

multi-aged stands of trees that exhibited high levels of structural complexity. These sites are generally on flat terrain and lower south-facing slopes which have horizon shading and lower levels of solar radiation (Lindenmayer, chapter 4). Similarly in Scandinavia (Niemela, chapter 7), cutting regimes can be customised to particular vegetative communities with a landscape of varying fire frequencies and intensities.[2] Harvesting methods that offer alternatives to traditional clearcutting are a significant component of the transitions to ecological sustainability in Sweden (Angelstam, chapter 9), Finland (Niemela, chapter 7), and British Columbia (Bunnell & Kremsater, chapter 6). They are also being tested in Tasmania in preparation for implementation on an operational basis (Hickey & Brown, chapter 3). In Victoria, alternative methods of harvesting montane ash forests were examined as a part of the Silvicultural System Project but none have been widely adopted; furthermore, since all the treatments were adaptations of traditional silvicultural approaches, none included permanent retention of any

2 The ASIO model—see Rulcker et al 1994.

structural elements of the harvested stand. Some new integrated experimental and operational research has recently been established in montane ash forests, however (Lindenmayer, chapter 4). Ecologically sensitive approaches to harvesting native forests in New Zealand were being successfully tested prior to the government policy decision ending all such harvests (Norton, chapter 10).

Transitions to sustainability—learning as we go

Although the effectiveness of alternative silvicultural systems in meeting a range of conservation objectives is not yet fully known, a number of important research projects are underway. There are several important issues to be considered in such evaluations. First, an effective monitoring program is essential to determine whether an approach is effective. Despite its importance, the record of forest monitoring is appalling virtually everywhere (Dovers, chapter 2). It is incredible that the concept of thorough monitoring and assessment in natural resource management is not fully embraced even by large corporations that do have mechanisms to regularly assess their economic performance (even if it is not always reported faithfully to stockholders). It is critical that we develop mechanisms to insure that monitoring is adequate to the task, rigorous in execution, and assured of long-term implementation to better guide sustainable resource management and the fulfilment of obligations, such as those under forest certification agreements.

Continued resourcing of management and research efforts is a second fundamental issue in transitioning to ecological sustainable management. Unfortunately, the intellectual capital of forest management agencies has been seriously depleted in many jurisdictions; this obviously needs to be corrected. Without monitoring and research, which often can and should be difficult to distinguish, forestry organisations can **not** claim to have embraced adaptive management. Under an adaptive management strategy, forestry practices are continuously improved on the basis of new knowledge; it is a key indicator that an organisation has embraced ecologically sustainable forest management. Dovers (chapter 2) further notes that there should be **both** ecological and policy monitoring—the latter being an activity that has been very uncommon.

Avoiding monocultures of the mind—and the land

It seems unlikely—based on the preceding discussion—that partitioning the forest estate into plantations (fibre farms) and reserves will achieve all conservation objectives and resolve the remaining conflicts between ecological and commodity goals. Such a simplistic approach is an example of what Shiva (1993) describes as 'monocultures of the mind'.[3] Realistically, such simple approaches ignore the real complexities inherent in such complex entities as forest ecosystem and human societies and, of course, their interrelationship.[4]

Reserve and off-reserve management will be necessary, therefore, in order to achieve the goals of conserving biodiversity and maintaining critical ecosystem functions. This applies even in those regions and countries where it is feasible to move forest harvesting completely out of native forests and into plantations. There will still be important management activities that will have to be conducted in these forests to meet other objectives, even highly focussed ecological and biodiversity objectives (see Franklin, chapter 1). Similarly, there will be significant environmental and social issues in areas broadly designated as plantations that will require management for values other than strict wood production. This is the complex plantation model proposed by Kanowski.[5]

The future

Predicting the future is invariably an exercise in futility—what eventually transpires is rarely congruent with predictions! However, what does seem certain is that there will continue to be conflict over the management and conservation of forests in most developed nations. This continuing conflict is based on three key factors (Bunnell & Kremsater, chapter 6): an educated public, an affluent public, and a resource worth fighting about. Dovers (chapter 2) proposes some additional ingredients. First, forest trees are long-lived plants, a trait that does not fit well with the short-

---

3 Shiva 1993.

4 The need for a more complex and comprehensive approach in forest conservation is discussed in Lindenmayer & Franklin 2002.

5 Kanowski 1997.

term perspectives of capitalistic economies (and short-term managerialsism). Second, some types of forest, such as those on high productivity sites or those that are old, are relatively scarce and humans tend to value such scarce objects highly.

It should not be surprising then, that all the examples provided in this book of transitions to ecological sustainability have been built on a base of prolonged public debate over forest policy. It is often the case that the highly polarised nature of these debates has impaired progress (Niemela, chapter 7) and good planning (Bunnell & Kremsater, chapter 6) or both.

Reflecting this, one of the outcomes of the Forestry Roundtable Meeting at Marysville was a request to make all of the information used in forest planning and decision-making available to all stakeholder groups. This will lead to greater participatory (rather than consultative) decision-making, increase the transparency of the policy development process, and contribute to conflict reduction.

In summary, ecological sustainability is not an endpoint but, rather, an **overall direction in conservation and forest management**. There will be many transitions and not all the movement will be forward! Forward movement is essential but it will not always be easy. As several participants in forest debates have noted, the challenges associated with ecologically sustainable forest management are not rocket science—they are much harder! (Bunnell & Kremsater, chapter 6). However, given the stakes, these challenge cannot be avoided.

References and bibliography

Agee, J.K. 1993. *Fire Ecology of Pacific Northwest Forests.* Island Press: Washington DC.

Andrén, H. 1994. Effects of habitat fragmentation on birds and mammals in landscapes with different proportions of suitable habitat: a review. *Oikos* **71**, 355–366.

Angelstam, P. 1996. Ghost of forest past – natural disturbance regimes as a basis for reconstruction of biologically diverse forests in Europe In *Conservation of Faunal Diversity in Forested Landscapes.* Eds R. DeGraaf, & R.L. Miller. pp. 287–337.Chapman & Hall: London

Angelstam, P. 1997. Landscape analysis as a tool for the scientific management of biodiversity. *Ecological Bulletin* **46**, 140–70.

Angelstam, P. 1998. Maintaining and restoring biodiversity by developing natural disturbance regimes in European boreal forest. *Journal of Vegetation Science* **9**, 593–602.

Angelstam, P. In press. Reconciling land management with natural disturbance regimes for the maintenance of forest biodiversity in Europe. In *Landscape ecology and resource management: Linking theory with practice.* Eds J. Bissonette & I. Storch. Island Press: Washington, DC.

Angelstam, P. & Andersson, L. 2001. Estimates of the needs for forest reserves in Sweden. *Scandinavian Journal of Forest Research* **16**, supplement 3, 38–51.

Angelstam, P., Anufriev, V., Balciauskas, L., Blagovidov, A., Borgegård, S O., Hodge, S., Majewski, P., Shvarts, E., Tishkov, A , Tomialojc, L & Wesolowski, L. 1997. Biodiversity and sustainable forestry in European forests – how west and east can learn from each other. *Wildlife Society Bulletin* **25**(1), 38–48.

Angelstam, P. & Breuss, M. (eds) 2001. Critical habitat thresholds & monitoring tools for the practical assessment of forest biodiversity in boreal forest. Report to MISTRA. http://iufro.boku.ac.at/iufro/iufronet/d8/hp80206.htm

Angelstam, P., Breuss, M. & Mikusinski, G. 2001. Toward the assessment of forest biodiversity of forest management units – a European perspective. In Criteria and indicators for sustainable forest management at the forest management unit level. Eds A. Franc, O. Laroussinie & T. Karjalainen. *European Institute Proceedings* **38**, 59–74. Gummerus printing: Saarijärvi, Finland.

Angelstam, P. & Pettersson, B. 1997. Principles of present Swedish forest biodiversity management. *Ecological Bulletins* **46**, 191–203.

Ashton, D.H. 1976. The development of even-aged stands of *Eucalyptus regnans* F. Muell. in central Victoria. *Australian Journal of Botany* **24**, 397–414.

Ashton, D.H. 1986. Ecology of bryophytic communities in mature *Eucalyptus regnans* F. Muell. forest at Wallaby Creek, Victoria. *Australian Journal of Botany* **34**, 107–29.

Attiwill, P.M. 1994. Ecological disturbance and the conservative management of eucalypt forests in Australia. *Forest Ecology and Management* **63**, 301–46.

Ball, I., Lindenmayer, D.B. and Possingham, H.P. 1999. HOLSIM: a model for simulating hollow availability in managed forest stands. *Forest Ecology and Management* **123**, 179–94.

Beese, W.J. & Bryant, A.A. 1999. Effect of alternative silvicultural systems on vegetation and bird communities in coastal montane forests of British Columbia, Canada. *Forest Ecology and Management* **115**, 231–42.

Beese, W.J., Dunsworth, G. & Perry, J. 2001. The Forest Project: three year review and update. *Ecoforestry* **16**(4), 10–17.

Boland, D.J., Brooker, M.I., Chippendale, G.M., Hall, N., Hyland, B.P., Johnston, R.D., Kleinig, D.A. & Turner, J.D. 1984. *Forest trees of Australia*. CSIRO Publishing: Melbourne.

Borschmann G. 1999. *The people's forests: A living history of the Australian bush*. The People's Forest Press: Blackheath NSW.

Brown, G.W. & Nelson, J.L. 1993. Influence of successional stage of *Eucalyptus regnans* (Mountain Ash) on habitat use by reptiles in the Central Highlands of Victoria. *Australian Journal of Ecology* **18**, 405–18.

Brown, G.W., Nelson, J.L. & Cherry, K.A. 1997. The influence of habitat structure on insectivorous bat activity in montane ash forests of the Central Highlands of Victoria. *Australian Forestry* **60**, 138–146.

Brown, M.J. 1996. Benign neglect and active management in Tasmania's forests: a dynamic balance or ecological collapse. *Forest Ecology and Management* **85**, 279–89.

Brown, M.J., Elliott, H.J. & Hickey, J.E. 2001. An overview of the Warra long-term ecological research site. *Tasforests* **13**(1), 1–8.

Bunnell, F.L. 1998. Time and space without Einstein: How do we distribute our treatments? *Ecoforestry* **13**(3), 33–9.

Bunnell, F.L., Dunsworth, B.G., Huggard, D.J. & Kremsater, L.L. 2003. *Learning to sustain biological diversity on Weyerhaeuser's coastal tenure*. chs 1–6. Weyerhaeuser: Nanaimo BC.

Bunnell, F.L. & Johnson J.F. (eds). 1998. *The living dance: Policy and practices for biodiversity in managed forests*. University of British Columbia Press: Vancouver.

Bunnell, F.L., Kremsater, L.L. & Wind, E. 1999. Managing to sustain vertebrate richness in forests of the Pacific Northwest: Relationships within stands. *Environmental Reviews* **7**, 97–146.

Campbell, R. 1997. Evaluation and development of sustainable silvicultural systems for multiple-purpose management of mountain ash forests: discussion paper. VSP Technical Report No. 28. Natural Resources and Environment: Melbourne.

Campbell, R.G. (compiler). 1997. Evaluation and development of sustainable silvicultural systems for Mountain Ash forests. Discussion paper. Value Adding and Silvicultural Systems Report. VSP Technical Report No. 28. Forests Service, Department of Natural Resources and Environment: Melbourne. July 1997.

Carey, A.B. 1993. The forest ecosystem study: Experimental manipulation of managed stands to provide habitat for Spotted Owls and to enhance plant and animal diversity: A summary and background for the interagency experiment at Fort Lewis, Washington. Forestry Sciences Laboratory: Olympia, Washington.

Carey, A.B., Lippke, B.R. & Sessions, J. 1999. Intentional systems management: managing forests for biodiversity. *Journal of Sustainable Forestry* **9**(3/4), 83–125.

Carey, A.B. & Sanderson, H.R. 1981. Routine to accelerate tree cavity formation. *Wildlife Society Bulletin* **9**, 14–21.

Carlsson, M. 1998. *On forestry planning for timber and biodiversity – the landscape perspective*. Acta Universitatis Agriculturae Sueciae. Silvestria 64, Alnarp.

Churton, N.L., Bren, L.J. & Kerruish, C.M. 1996. Efficiency of mechanism Mountain Ash thinning in the Central Highlands of Victoria. *Australian Forestry* **62**, 72–8.

Clark J. 2002. A new forest and wood industry policy framework for Australia. Paper presented at the Australia New Zealand Society for Ecological Economics conference *Strategies into Action: Regional and Industry Policy Applications of Ecologically Sustainable Development*. 2–4 December, University of Technology, Sydney, Australia.

Clark J. 1995. *Australia's plantations: Industry, employment, environment.* Report to the State Conservation Councils of Australia. Melbourne, Vic, Australia: Environment Victoria.

Clark J. 2001. The global wood market, prices and plantation investment: an examination drawing on the Australian experience. *Environmental Conservation* **28** (1), 53–64.

Commonwealth of Australia and Department of Natural Resources and Environment. 1997. Comprehensive Regional Assessment – Biodiversity. Central Highlands of Victoria. The Commonwealth of Australia and Department of Natural Resources and Environment: Canberra.

Commonwealth of Australia and State of Tasmania 1997. *Tasmanian Regional Forest Agreement between the Commonwealth of Australia and the State of Tasmania.*

Costanza R. & Patten B.C. 1995. Defining and predicting sustainability, *Ecological Economics* **15**, 193–6.

Crabb, P. 2003. Straddling boundaries: inter-governmental arrangements for managing natural resources. In *Managing Australia's environment.* Eds S. Dovers & S.Wild River. pp. 229–254. Federation Press: Sydney.

Crooks, K.R. & M.E. Soulé. 1999. Mesopredator release and avifaunal extinctions in a fragmented system. *Nature* **400**, 563–66.

Curtis, A. 2003. The Landcare experience. In *Managing Australia's environment.* Eds S. Dovers & S. Wild River. pp. 442–460. Federation Press: Sydney.

Dargarvel, J. 1995. *Fashioning Australia's forests.* Oxford University Press: Melbourne.

Dempster, K. 1962. Internal memorandum to Director (of Fisheries and Wildlife Division). October 2 1967. Fisheries and Wildlife Division: Melbourne.

Department of Conservation, Forests and Lands. 1989. Code of Practice. Code of Forest Practices for timber production. Revision No. 1, May 1989. Department of Conservation, Forests and Lands: Melbourne.

Department of Natural Resources and Environment. 1996. Code of Practice. Code of Forest Practices for timber production. Revision No. 2, November 1996. Department of Natural Resources and Environment: Melbourne.

Dore, J., Woodhill, J., Andrews, K. & Keating, C. 2002. *Sustainable regional development: lessons from Australian efforts.* Greening Australia: Canberra.

Dovers, S. 2001. *Institutions for sustainability.* Tela paper 7. Australian Conservation Foundation: Melbourne.

Dovers, S. 2002. Sustainability: reviewing Australia's progress, 1992–2002. *International Journal of Environmental Studies* **59**, 559–71.

Dovers, S. & Lindenmayer, D.B. 1997. Managing the environment: Rhetoric, policy and reality. *Australian Journal of Public Administration* **56**, 65–80.

Dovers, S. & Wild River, S. (eds). 2003. *Managing Australia's environment.* Federation Press: Sydney.

Duinker, P. 2001. Criteria and indicators of sustainable forest management in Canada: progress and problems in integrating science and politics at the local level. In Criteria and indicators for sustainable forest management at the forest management unit level. Eds A. Franc, O. Laroussinie & T. Karjalainen. *European Institute Proceedings* **38**, 7–37. Gummerus printing: Saarijärvi, Finland.

Dunsworth, G. & Beese, B. 2000. New approaches in managing temperate rainforests. In *Mountain forests and sustainable development*, prepared for The Commission on Sustainable Development (CSD) and its 2000 spring session by Mountain Agenda. pp. 24–25. Swiss Agency for Development and Cooperation, Berne, Switzerland.

Ellenberg, H. 1996. *Vegetation Mitteleuropas mit den Alpen.* 5. Auflage. Verlag Eugen Ulmer: Stuttgart.

Elliott, C. 1999. Forest certification: analysis from a policy network perspective. PhD thesis 1965. Ecole Polytechnique Federale de Lausanne. 464 pp.

Elliott, C., & Schlaepfer, R. 2001. Understanding forest certification using the advocacy coalition framework. *Forest Policy and Economics* **2**, 257–66.

Environmental Resources Management Australia. 2001. Environmental Impact Statement. Wood processing and metallurgical carbon facility for Australian Silicon Operations Pty Ltd. Environmental Resources Management Australia, November 2001.

Fahrig, L. 1999. Forest loss or fragmentation: which has the greater effect on the persistence of forest-dwelling animals? In *Forest fragmentation: Wildlife and management implications*. Eds J.A. Rochelle, L.A. Lehmann & J. Wisniewski. pp. 87–95. Brill: Boston.

Fahrig, L. 2001. How much is enough? *Biological Conservation* **100**, 65–74.

Ferguson, I.S., Fox, J., Baker, T., Stackpole, D. & Wild, I. 2002. *Plantations of Australia: Wood availability 2001–2044.* Bureau of Rural Sciences: Canberra.

Fischer, J. & Lindenmayer, D.B. 2002. The conservation value of small habitat patches: two case studies on birds from southeastern Australia. *Biological Conservation* **106**, 129–36.

Flannery, T. 2001. *The eternal frontier: An ecological history of North America and its peoples.* Atlantic Monthly Press: New York.

Flannery, T. 2002. *The future eaters: An ecological history of the Australasian lands and people.* Grove Press: New York.

Food and Agricultural Organisation of the United Nations 1997. *FAO Provisional Outlook for Global Forest Products Consumption, Production and Trade to 2010.* Food and Agricultural Organisation of the United Nations: Rome.

Forest Practices Board 2000. *Forest Practices Code.* Forest Practices Board: Hobart.

Forests and Forest Industry Council (1990). *Secure futures for forests and people.* Forests and Forest Industry Council: Hobart.

Franklin, J.F. 2000. Threads of continuity. *Conservation in Practice* **1**, 9–16

Franklin, J.F., Berg, D.R., Thornburgh, D.A. & Tappeiner, J.C. 1997. Alternative silvicultural approaches to timber harvesting: variable retention harvest systems. In *Creating a forestry for the 21st century: The science of ecosystem management.* Eds K.A. Kohm & J.F. Franklin. pp. 111–39. Island Press: Washington DC.

Fries, C., Johansson, O., Petterson, B. & Simonsson, P. 1997. Silvicultural models to maintain and restore natural stand structures in Swedish boreal forests. *Forest Ecology and Management* **94**, 89–103.

Gibbons, P. & Lindenmayer, D.B. 1997. Conserving hollow-dependent fauna in timber-production forests New South Wales. National Parks and Wildlife Service Environmental Heritage Monograph Series **3**, 1–110.

Gibbons, P. & Lindenmayer, D.B. 2002. *Tree hollows and wildlife conservation in Australia.* CSIRO Publishing, Melbourne.

Gilbert, J.M. 1959. Forest succession in the Florentine Valley, Tasmania. *Papers and Proceedings of the Royal Society of Tasmania,* **93**, 129–51.

Gooday, P., Whish-Wilson, P. & Weston, L. 1997. Regional Forest Agreements. Central Highlands of Victoria. pp. 1–11 Australian forest products statistics. Australian Bureau of Agricultural and Resource Economics: Canberra. September Quarter 1997.

Government of Victoria. 1988. Flora and Fauna Guarantee Act. No. 47 of 1988. Government Printer, Melbourne.

Government of Victoria. 1992. Flora and Fauna Guarantee Strategy: Conservation of Victoria's Biodiversity. Department of Conservation and Environment: Melbourne.

Government of Victoria. 1986. Government Statement No. **9**. Timber Industry Strategy. Government Printer, Melbourne.

Griffiths, A.D. 2002. Indigenous forestry on private land: present trends and future potential. Ministry of Agriculture and Forestry Technical Paper 01/6, Wellington.

Hamilton, C. 2003. The Resource Assessment Commission: lessons in the venality of modern politics. In *Managing Australia's environment*. Eds S. Dovers & S. Wild River. pp. 117–132. Federation Press: Sydney.

Hammond, P.C. & Miller, J.C. 1998. Comparison of the biodiversity of Lepidoptera within three forested ecosytems. *Annals of the Entomological Society of America* **91**, 323–328.

Hannah, L., Carr, J.L. & Lankerani, A. 1995. Human disturbance and natural habitat: a biome level analysis of a global data set. *Biodiversity and Conservation* **4**, 128–55.

Hanski, I. 2000. Extinction debt and species credit in boreal forests: modelling the consequences of different approaches to biodiversity conservation. *Annales Zoologici Fennici* **37**, 271–80.

Harris, L.D. 1984. *The fragmented forest: Island biogeography theory and the preservation of biotic diversity*. University of Chicago Press: Chicago.

Hartley, M.J. 2002. Rationale and methods for conserving biodiversity in plantation forests. *Forest Ecology and Management* **155**, 81–95.

The Heinz Center. 1999. Designing a report on the state of the nation's ecosystems: Selected measurements for croplands, forests and coasts & oceans. The H. John Heinz III Center. Washington, DC.

Hellström, E. & Reunala, A. 1995. Forestry conflicts from the 1950s to 1983. European Forest Institute Research Report 3.

Hickey, J.E., Neyland, M.G. & Bassett, O.D. 2001. Rationale and design for the Warra Silvicultural Systems trial in wet *Eucalyptus obliqua* forests in Tasmania. *Tasforests* **13**(2), 155–82.

Hickey, J.E. & Wilkinson, G.R. 1999. The development and current implementation of silvicultural practices in native forests in Tasmania. *Australian Forestry* **62**(3), 245–54.

Holling, C.S. (ed.) 1978. *Adaptive environmental assessment and management. international series on applied systems analysis 3*. International Institute for Applied Systems Analysis. John Wiley & Sons: Toronto.

Howard, T.M., 1973. Studies in the ecology of *Nothofagus cunninghamii* Oerst. I. Natural regeneration on the Mt. Donna Buang massif, Victoria. *Australian Journal of Botany* **21**, 67–78.

Humphreys C. 1992. The basic principles of marketing plantation logs. Proceedings *Commercial opportunities for private tree growing*, Australian Forest Growers biennial conference, Mount Buffalo, Victoria, Australia, 10–12 April 1992.

Hunter, M.L. 1994. *Fundamentals of Conservation Biology*. Blackwell: Cambridge.

Hunter, M.L. (ed.) 1999. *Maintaining biodiversity in forest ecosystems*. Cambridge University Press: Cambridge.

Incoll, R.D., Loyn, R.H., Ward, S.J., Cunningham, R.B. & Donnelly, C.F. 2000. The occurrence of gliding possums in old-growth patches of Mountain Ash (*Eucalyptus regnans*) in the Central Highlands of Victoria. *Biological Conservation* **98**, 77–88

Jackson, W.D. 1968. Fire, air, water and earth – an elemental ecology of Tasmania. *Proceedings of the Ecological Society of Australia* **3**, 9–16.

James, I.L. 1987. Silvicultural management of the rimu forests of south Westland. Forest Research Institute: Rotorua.

James, I.L., & Norton, D.A. 2001. Helicopter based natural forest management for New Zealand's rimu (*Dacrydium cupressinum*, Podocarpaceae) forests. *Forest Ecology and Management* **155**, 337–46.

Johnson, T.G. (ed). 1997. US Timber Industry–An Assessment of Timber Product Output and Use. USFS Gen. Teach. rep. SRS-45: Asheville, NC.

Jonsson, B.G. & Kruys, N. (eds) 2001. Ecology of woody debris in boreal forests. *Ecological Bulletins* **49**.

Kanowski, P. 1997. Plantation forestry in the 21st century. Special paper: afforestation and plantation forestry. XI World Forestry Congress, 13–22 October 1997.

Korpilahti, E. & Kuuluvainen, T. 2002. Disturbance dynamics in boreal forests: defining the ecological basis of restoration and management of biodiversity. *Silva Fennica* **36**(1).

Land Conservation Council. 1994. Final recommendations. Melbourne Area. District 2 Review. Land Conservation Council: Melbourne.

Larsson, S. & Danell, K. (eds) 2001. Science and the management of boreal forest biodiversity. *Scandinavian Journal of Forest Research* **16** supplement 3.

Larsson, T-B. (ed). 2001. Biodiversity evaluation tools for European forests. *Ecological Bulletins* **50**.

Lee, K.N. 1993. *Compass and Gyroscope: Integrating science and policy for the environment.* Island Press: Washington DC.

Leslie A.J. 2001. The uncompromising future. *Unasylva* **52** (1), 6–7.

Liaison Unit in Lisbon. 1998. Third ministerial conference on the protection of forests in Europe. General declarations and resolutions adopted. Ministry of Agriculture: Lisbon.

Lindenmayer, D.B. 1997. Differences in the biology and ecology of arboreal marsupials in southeastern Australian forests and some implications for conservation. *Journal of Mammalogy* **78**, 1117 1127.

Lindenmayer, D.B. 1992. The ecology and habitat requirements of arboreal marsupials in the montane ash forests of the Central Highlands of Victoria. A summary of studies. Value Adding and Silvicultural Systems Program, No. 6. Native Forest Research. Department of Conservation and Environment: Melbourne.

Lindenmayer, D.B. 1989. The ecology and habitat requirements of Leadbeater's possum. PhD thesis, Australian National University: Canberra.

Lindenmayer, D.B. 2000. Factors at multiple scales affecting distribution patterns and its implications for animal conservation – Leadbeater's possum as a case study. *Biodiversity and Conservation* **9**, 15–35.

Lindenmayer, D.B. 1996. *Wildlife and woodchips: Leadbeater's possum as a testcase of sustainable forestry.* University of New South Wales Press: Sydney.

Lindenmayer, D., Claridge, A.W., Gilmore, A.M., Michael, D., & Lindenmayer, B.D. 2002. The ecological roles of logs in Australian forests and the potential impacts of harvesting intensification on log-using biota. *Pacific Conservation Biology* **8**(2), 121-140.

Lindenmayer, D.B. & Cunningham, R.B. 1996. A habitat-based microscale forest classification system for zoning wood production areas to conserve a rare species threatened by logging operations in south-eastern Australia. *Environmental Monitoring and Assessment* **39**, 543–57.

Lindenmayer, D.B., Cunningham, R.B. & Donnelly, C.F. 1994b. The conservation of arboreal marsupials in the montane ash forests of the Central Highlands of Victoria, south-east Australia. VI. Tests of the performance of models of nest tree and habitat requirements of arboreal marsupials. *Biological Conservation* **70**, 143–147.

Lindenmayer, D.B., Cunningham, R.B. & Donnelly, C.F. 1997. Tree decline and collapse in Australian forests: implications for arboreal marsupials. *Ecological Applications* **7**, 625–641.

Lindenmayer, D.B., Cunningham, R.B., Donnelly, C.F. & Franklin, J.F. 2000a. Structural features of Australian old-growth montane ash forests. *Forest Ecology and Management* **134**, 189–204.

Lindenmayer, D.B., Cunningham, R.B., Donnelly, C.F., Nix, H.A. & Lindenmayer, B.D. 2002. The distribution of birds in a novel landscape context. *Ecological Monographs* **72**, 1–18.

Lindenmayer, D.B., Cunningham, R.B., Donnelly, C.F., Tanton, M.T. & Nix, H.A. 1993a. The abundance and development of cavities in montane ash-type eucalypt trees in the montane forests of the Central Highlands of Victoria, south-eastern Australia. *Forest Ecology and Management* **60**, 77–104.

Lindenmayer, D.B., Cunningham, R.B., Donnelly, C.F., Tanton, M.T. & Nix, H.A. 1993b. The conservation of arboreal marsupials in the montane ash forests of the Central Highlands of Victoria, south-east Australia. IV. The distribution and abundance of arboreal marsupials in retained linear strips (wildlife corridors) in timber production forests. *Biological Conservation* **66**, 207–21.

Lindenmayer, D.B., Cunningham, R.B., Donnelly, C.F., Triggs, B.J. & Belvedere, M. 1994a. The conservation of arboreal marsupials in the montane ash forests of the Central Highlands of Victoria, south-east Australia. V. Patterns of use and the microhabitat requirements of the Mountain Brushtail Possum, *Trichosurus caninus* Ogilby in retained linear strips (wildlife corridors). *Biological Conservation* **68**, 43–51.

Lindenmayer, D.B., Cunningham, R.B. & McCarthy, M.A. 1999a. The conservation of arboreal marsupials in the montane ash forests of the Central Highlands of Victoria, south-eastern Australia. VIII. Landscape analysis of the occurrence of arboreal marsupials in the montane ash forests. *Biological Conservation* **89**, 83–92.

Lindenmayer, D.B., Cunningham, R.B., Nix, H.A., Tanton, M.T. & Smith, A.P. 1991c. Characteristics of hollow-bearing trees occupied by arboreal marsupials in the montane ash forests of the Central Highlands of Victoria, south-eastern Australia. *Forest Ecology and Management* **40**, 289–308.

Lindenmayer, D.B., Cunningham, R.B., Nix, H.A., Tanton, M.T. & Smith, A.P. 1991b. The conservation of arboreal marsupials in the montane ash forests of the Central Highlands of Victoria, south-east Australia. III. The habitat requirements of Leadbeater's possum, *Gymnobelideus leadbeateri* McCoy and models of the diversity and abundance of arboreal marsupials. *Biological Conservation* **56**, 295–315.

Lindenmayer, D.B., Cunningham, R.B., Nix, H.A., Tanton, M.T. & Smith, A.P. 1991a. Predicting the abundance of hollow-bearing trees in montane ash forests of south-eastern Australia. *Australian Journal of Ecology* **16**, 91–98.

Lindenmayer, D.B., Cunningham, R.B., Tanton, M.T. & Smith, A.P. 1990. The conservation of arboreal marsupials in the montane ash forests of the Central Highlands of Victoria, south-east Australia. II. The loss of trees with hollows and its implications for the conservation of Leadbeaters Possum *Gymnobelideus leadbeateri* McCoy (Marsupialia: Petauridae). *Biological Conservation* **54**, 133–45.

Lindenmayer, D.B. & Franklin, J.F. 2002. *Conserving forest biodiversity: A comprehensive multiscaled approach.* Island Press: Washington DC.

Lindenmayer, D.B. & Franklin, J.F. 1997. Forest structure and sustainable temperate forestry: A case study from Australia. *Conservation Biology* **11**, 1053–68.

Lindenmayer, D.B. & Incoll, R.D.1998. Community-based monitoring of vertebrates in Victorian forests. *On the Brink* **11**, 11 12.

Lindenmayer, D.B., Incoll, R.D., Cunningham, R.B & Donnelly, C.F. 1999c. Attributes of logs in the Mountain Ash forests in south-eastern Australia. *Forest Ecology and Management* **123**, 195–203.

Lindenmayer, D.B. & Lacy, R.C. 1995. Metapopulation viability of arboreal marsupials in fragmented old-growth forests: a comparison among species. *Ecological Applications* **5**, 183–199.

Lindenmayer, D.B., Mackey, B., Mullins, I., McCarthy, M.A., Gill, A.M., Cunningham, R.B. & Donnelly, C.F. 1999b. Stand structure within forest types – are there environmental determinants? *Forest Ecology and Management* **123**, 55–63.

Lindenmayer, D.B., Mackey, B.G., Cunningham, R.B., Donnelly, C.F., Mullen, I., McCarthy, M.A & Gill, A.M. 2000b. Statistical and Environmental Modelling of Myrtle Beech (*Nothofagus cunninghamii*) in southern Australia. *Journal of Biogeography* **27**, 1001–9.

Lindenmayer, D. & McCarthy, M.A. 2002. Congruence between natural and human forest disturbance: a case study from Australian montane ash forests. *Forest Ecology and Management* **155**, 319-35.

Lindenmayer, D.B.,& Nix, H.A. 1993. Ecological principles for the design of wildlife corridors. *Conservation Biology* **7**, 627–30.

Lindenmayer, D.B. & Norton, T.W. 1993. The conservation of Leadbeater's possum in south-eastern Australia and the Spotted Owl in the Pacific Northwest of the U.S.A.: management issues, strategies and lessons. *Pacific Conservation Biology* **1**, 13–18.

Lindenmayer, D.B. & Possingham, H.P. 1995. Modeling the impacts of wildfire on metapopulation behaviour of the Australian arboreal marsupial, Leadbeater's possum, *Gymnobelideus leadbeateri*. *Forest Ecology and Management* **74**, 197–222.

Lindenmayer, D.B. & Possingham, H.P. 1996. Modeling the relationships between habitat connectivity, corridor design and wildlife conservation within intensively logged wood production forests of south-eastern Australia. *Landscape Ecology* **11**, 79–105.

Lindenmayer, D.B. & Possingham, H.P. 1994. *The risk of extinction: ranking management options for Leadbeater's possum*. Centre for Resource and Environmental Studies, The Australian National University and The Australian Nature Conservation Agency: Canberra.

Loyn, R.H. 1985. Bird populations in successional forests of Mountain Ash *Eucalyptus regnans* in Central Victoria. *Emu* **85**, 213–30.

Loyn, R.H. 1998. Birds in patches of old-growth ash forest, in a matrix of younger forest. *Pacific Conservation Biology* **4**, 111–21.

Lumsden, L.F., Alexander, J.S.A., Hill, F.A.R., Krasna, S.P. & Silveira, C.E. 1991. The vertebrate fauna of the Land Conservation Council Melbourne–2 study area. Arthur Rylah Institute for Environmental Research. Technical Report Series No. **115**. Department of Conservation and Environment: Melbourne.

Lutze, M.T., Campbell, R.G. & Fagg, P.C. 1999. Development of silviculture in the native state forests of Victoria. *Australian Forestry* **62**, 236–244.

Macfarlane, M.A. & Seebeck, J.H.1991. Draft management strategies for the conservation of Leadbeater's Possum, *Gymnobelideus leadbeateri*, in Victoria. Arthur Rylah Institute Technical Report Series **111**. Department of Conservation and Environment: Melbourne.

Macfarlane, M.A., Smith, J. & Lowe, K. 1998. Leadbeater's Possum Recovery Plan. 1998–2002. Department of Natural Resources and Environment. Government of Victoria.

Mackey, B.G., Lindenmayer, D.B., Gill, A.M. & McCarthy, M.A. 2002. *Wildlife, fire and future climate: A forest ecosystem analysis.* CSIRO Publishing: Melbourne.

MAF 2002. *Standards and guidelines for the sustainable management of indigenous forests.* Ministry of Agriculture and Forestry: Wellington.

Marchak, M. Patricia. 1995. *Logging the globe.* McGill-Queen's University Press: Montreal.

Marsden, S. & Dovers, S. (eds) 2002. *Strategic environmental assessment in Australia.* Federation Press: Sydney.

Martin, P.S. & Klein, R.G. (eds). 1984. *Quarternary extinctions: A prehistoric revolution.* University of Arizona Press: Tucson.

Mather, A.S. 1990. *Global forest resources.* Belhaven Press: London.

McCarthy, M.A., Gill, A.M. & Lindenmayer, D.B. 1999. Fire regimes in mountain ash forest: evidence from forest age structure, extinction models and wildlife habitat. *Forest Ecology and Management* **124**, 193–203.

McCarthy, M.A. & Lindenmayer, D.B. 1998. Development of multi-aged mountain ash (*Eucalyptus regnans*) forest under natural disturbance regimes: implications for wildlife conservation and logging practices. *Forest Ecology and Management* **104**, 43–56.

McCarthy, M.A. & Lindenmayer, D.B. 1999a. Incorporating metapopulation dynamics of Greater Gliders into reserve design in disturbed landscapes. *Ecology* **80**, 651–667.

McCarthy, M.A. & Lindenmayer, D.B. 1999b. Spatially correlated extinction in metapopulation models. *Biodiversity and Conservation* **9**, 47–63.

Milledge, D.R., Palmer, C.L. & Nelson, J.L. 1991. 'Barometers of change': the distribution of large owls and gliders in mountain ash forests of the Victorian Central Highlands and their potential as management indicators. In *Conservation of Australia's forest fauna.* Ed. D. Lunney. pp. 55–65. Royal Zoological Society of NSW: Sydney.

Mitchell, S.J. & Beese, W.J. 2002. The retention system: reconciling variable retention with the principles of silvicultural systems. *The Forestry Chronicle*, **78**, 397–403.

Mobbs, C. 2003. National forest policy and Regional Forest Agreements. In *Managing Australia's environment.* Eds S. Dovers and S. Wild River. pp. 90–114. Federation Press: Sydney.

Moore, S.E. & Allen, H.L. 1999. Plantation forestry. In *Managing biodiversity in forest ecosystems.* Ed. M.L. Hunter. pp. 400–433. Cambridge University Press: Cambridge.

Mueck, S.G. 1990. The floristic composition of mountain ash and alpine ash forests in Victoria. SSP Technical Report No. 4. Department of Conservation and Environment: Melbourne.

Mueck, S.G., Ough, K. & Banks, J.C. 1996. How old are wet forest understories? *Australian Journal of Ecology* **21**, 345–8.

National Research Council. 1990. *Forestry research: A mandate for change.* National Academy Press: Washington, DC.

Nelson, J.L. & Morris, B.J. 1994. Nesting requirements of the yellow-tailed black cockatoo *Calyptorynchus funereus* in mountain ash forest (*Eucalyptus regnans*) and implications for forest management. *Wildlife Research* **21**, 267–78.

Noble, W.S. 1977. *Ordeal by fire. The week a state burned up*. Hawthorn Press: Melbourne.

Norton, D.A. 1998. Indigenous biodiversity conservation and plantation forestry: options for the future. *New Zealand Forestry* **43**, 34–9.

Norton, D.A. & Miller, C.J. 2000. Some issues and options for the conservation of native biodiversity in rural New Zealand. *Ecological Management and Restoration* **1**, 26–34.

Noss, Reed F. & Peters, Robert L. 1995. *Endangered ecosystems: a status report on America's vanishing habitat and wildlife*. Defenders of Wildlife: Washington, DC.

O'Shaughnessy, P. & Jayasuriya. J. 1991. In *Managing the ash-type forest for water production in Victoria: Forest Management in Australia*. Eds F.H. McKinnell, E.R. Hopkins & J.E.D. Fox. pp. 341–63. Surrey Beatty and Sons: Chipping Norton.

Ough, K. & Murphy, A. 1996. The effect of clearfell logging on tree-ferns in Victorian wet forest. *Australian Forestry* **59**, 178–88.

Ough, K. & Murphy, A. 1998. Understorey islands: a method of protecting understorey flora during clearfelling operations. Department of Natural Resources and Environment. Internal VSP Report No. 29. Department of Natural Resources and Environment: Melbourne.

Ough, K. & Ross, J. 1992. Floristics, fire and clearfelling in wet forests of the Central Highlands of Victoria. Silvicultural Systems Project Technical Report No. 11. Department of Conservation and Environment: Melbourne.

Perry, D.A. 1994. *Forest ecosystems*. The Johns Hopkins University Press: Baltimore.

Perry, D.A. 1998. The scientific basis of forestry. *Annual Review of Ecology and Systematics* **29**, 435-66.

Peterken, G. 1996. *Natural woodland. Ecology and conservation in northern temperate regions*. Cambridge University Press: Cambridge.

Pittock, J. 1994. Logging abuse. *Parkwatch* **31**, 6–8.

Quigley, T.M., Haynes R.W. & Graham R.T. 1996. Integrated scientific assessment for ecosystem management in the interior Columbia Basin and portions of the Klamath and Great Basins. USDA Forest Service General Technical Report PNW-GTR-282.

Rab, M.A. 1998. Rehabilitation of snig tracks and landings following logging of *Eucalyptus regnans* forests in the Victorian Central Highlands – a review. *Australian Forestry* **61**, 103–13.

Rawlinson, P.A. & Brown, P.R. 1980. Forestry practices threaten wildlife – the case for Leadbeater's possum. In *What state is the garden in?* Eds P. Westbrook & J. Farhall. pp. 57–61. Conservation Council Victoria: Melbourne.

Remmert, H. (ed.). 1991. The mosaic cycle concept of ecosystems. *Ecol. Stud. 85.* Springer-Verlag, Heidelberg.

Resource Planning and Development Commission 2002. Inquiry on the Progress with implementation of the Tasmanian Regional Forest Agreement (1997). Background Report. Resource Planning and Development Commission: Hobart.

Roche, M. 1990. *History of New Zealand forestry.* New Zealand Forestry Corporation: Wellington.

Rochelle J.A., Lehmann L.A. & Wisniewski, J. (eds). 1999. *Forest fragmentation: Wildlife and management implications.* Brill: Leiden, Netherlands.

Rosoman, G. 1994. *The plantation effect – an ecoforestry review of the environmental effects of exotic monoculture tree plantation in Aotearoa/New Zealand.* Greenpeace: Auckland.

Rülcker, C., Angelstam, P. & Rosenberg, P. 1994. Natural forest-fire dynamics can guide conservation and silviculture in boreal forests. *SkogForsk* **2**, 1–4.

Ruthven P. 1995. An endless war of attrition. *The Australian Financial Review*, 25 January 1995

Saveneh, A.G. & Dignan, P. 1998. The use of shelterwood in *Eucalyptus regnans* forest: the effect of overwood removal at three years on regeneration stocking and health. *Australian Forestry* **4**, 251–59.

Schoonmaker, P.K., von Hagen, B.& Wolf, E.C. 1997. Introduction. In *The rain forests of home: Profile of a North American bioregion.* Eds P.K. Schoonmaker, B. von Hagen & E.C. Wolf. pp.1–6. Island Press: Washington DC.

Sedjo R.A. 1983. *The comparative economics of plantation forestry.* Resources for the Future Inc: Washington DC.

Sedjo R.A. & Lyon K.S. 1990. *The long-term adequacy of world timber supply.* Resources for the Future: Washington, DC.

Shiva, V. 1993. *Monocultures of the mind.* Zed Books: London.

Sierra Nevada Ecosystem Project. 1996. Status of the Sierra Nevada. Volume II Assessments and scientific basis for management options. Final report to Congress. 1528 p. University of California Davis Wildland Resources Center Report No. 37.

Siitonen, J. 2001. Forest management, coarse woody debris and saproxylic organisms: Fennoscandian boreal forests as an example. *Ecological Bulletins* **49**, 11–41.

Smith, A.P. & Lindenmayer, D.B. 1992. Forest succession, timber production and conservation of Leadbeater's possum (*Gymnobelideus leadbeateri* Marsupialia: Petauridae). *Forest Ecology and Management* **49**, 311–32.

Smith, R.B. & Woodgate, P. 1985. Appraisal of fire damage for timber salvage by remote sensing in mountain ash forests. *Australian Forestry* **48**, 252–63.

Sohngen B., Mendelsohn R. & Sedjo R. 1999. Forest management, conservation and global timber markets. *American Journal of Agricultural Economics* **81** (1), 1–13.

Soulé, M.E., Estes, J.A., Berger, J. & Martinez del Rio, C. In press. Ecological effectiveness: conservation goals for interactive species. *Conservation Biology.*

Soulé, M.E. & Lease, G. (eds). *Reinventing nature? Responses to postmodern deconstruction.* Island Press: Washington DC.

Soulé, M.E. & Sanjayan, M. 1998. Conservation targets: Do they help? *Science* **279**, 2060–2061.

Soulé, M. E. & Terborgh, J. (eds). 1999. *Continental conservation: design and management principles for long-term, regional conservation networks.* Island Press: Washington DC.

Squire, R.O. 1993. The professional challenge of balancing sustained wood production and ecosystem conservation in the native forests of south-eastern Australia. *Australian Forestry* **56**, 237–48.

Squire, R.O. 1990. Report on the progress of the Silvicultural Systems Project July 1986–June 1989. Department of Conservation and Environment: Melbourne.

Squire, R.O. 1987. Revised treatments, design and implementation strategy for the Silvicultural Systems Project. Lands and Forests Division. Department of Conservation, Forests and Lands: Melbourne. August, 1987.

Squire, R.O., Campbell, R.G., Wareing, K.J. & Featherston, G.R. 1991. The mountain ash forests of Victoria: Ecology, silviculture and management for wood production. In *Forest management in Australia.*

Eds F.H. McKinnell, E.R. Hopkins & J.E.D. Fox. pp. 38–57. Surrey Beatty and Sons: Chipping Norton.

Squire, R.O., Dexter, B., Smith, R., Manderson, A. & Flinn, D. 1987. Evaluation of alternative systems for Victoria's commercially important mountain forests – project brief. Lands and Forests Division. Department of Conservation, Forests and Lands: Melbourne.

Stittholt, J.R., Noss, R.F., Frost, P.A., Vance-Borland, K., Carroll, C. & Hellman Jr, G. 1999. *A science-based conservation assessment for the Klamath-Siskiyou region.* Earth Design Consultants and the Conservation Biology Institute: Corvallis, Oregon.

Thomas, J.W. (ed). 1979. *Wildlife habitats of managed forests: the Blue Mountains of Oregon and Washington.* Agriculture Handbook No. 553. USFS: Washington DC.

USDA. 2000. Renewable Resources Planning Act Assessment. USDA: Washington, DC.

Van der Meer, P.J., Dignan, P. & Savaneh, A.G. 1999. Effects of gap size on seedling establishment, growth and survival at three years in mountain ash (*Eucalyptus regnans* F. Muell.) forest in Victoria, Australia. *Forest Ecology and Management* **117**, 33–42.

Voute, A.D. 1964. Harmonious control of forest insects. *International Review of Forest Research* **1**, 325–383.

Walker, K.J. 1992. Conclusion: the politics of environmental policy. In *Australian environmental policy: ten case studies.* Ed. K.J. Walker. University of NSW Press: Sydney.

Wardle, P. 1991. *Vegetation of New Zealand.* Cambridge University Press: Cambridge.

Warneke, R.M. 1962. Internal memorandum to Director (of Fisheries and Wildlife Division). May 10 1962. Fisheries and Wildlife Division: Melbourne.

Weyerhauser. 2000. Summary of the Second Year Critique Workshop on the Weyerhauser BC Coastal Forest Project. July 12–14, 2000. Weyerhauser: Vancouver.

Williams, J., Read, C., Norton, T., Dovers, S., Burgman, M., Proctor, W. & Anderson, H. 2001. *Australia state of environment 2001: Biodiversity theme report.* CSIRO Publishing: Melbourne.

Yaroshenko, A.Yu., Potapov, P.V. & Turubanova, S.A. 2001. *The intact forest landscapes of Northern European Russia.* Greenpeace Russia and the Global Forest Watch: Moscow.

Yencken, D. & Wilkinson, D. 2000. *Resetting the compass: Australia's journey towards sustainability.* CSIRO Publishing: Melbourne.